4BE

Wolfe Medical Publications Ltd

Titles in this series, published or being developed, include:
Diagnostic Picture Tests in Paediatrics
Picture Tests in Human Anatomy
Diagnostic Picture Tests in Oral Medicine
Diagnostic Picture Tests in Orthopaedics
Diagnostic Picture Tests in Infectious Diseases
Diagnostic Picture Tests in Dermatology
Diagnostic Picture Tests in Ophthalmology
Diagnostic Picture Tests in Rheumatology
Diagnostic Picture Tests in Obstetrics/Gynaecology
Diagnostic Picture Tests in Clinical Neurology
Diagnostic Picture Tests in Injury in Sport
Diagnostic Picture Tests in General Surgery
Diagnostic Picture Tests in General Medicine
Diagnostic Picture Tests in Paediatric Dentistry
Diagnostic Picture Tests in Dentistry
Picture Tests in Embryology

First published 1989 by Wolfe Medical Publications Ltd
Printed by W.S. Cowell Ltd, Ipswich, England
ISBN 0 7234 1519 6

For a full list of Wolfe Medical Atlases, plus
forthcoming titles and details of our surgical,
dental and veterinary Atlases, please write to
Wolfe Medical Publications Limited,
2-16 Torrington Place, London WC1E 7LT

A CIP catalogue record for this book
is available from the British Library.

Preface

Human embryology is a difficult subject to learn because of the dearth of specimens at many institutions. It is hoped that students who have never encountered human material will find the pictures useful. All of the photographs relate to human tissue.

The book is designed to help undergraduate medical students to revise for their examinations, and postgraduate medical students to review their basic embryology studies.

Within the limitations set by the length of this book it has not been possible to cover all subjects, nor is any single subject dealt with in depth. Rather, the questions are intended to serve as a guide to a student's overall comprehension of the subject and to serve as a stimulus to further reading and study.

Each illustration has an associated series of questions. The first question is related directly to the photograph, while subsequent ones may not necessarily refer directly to the illustration. This approach permits a wider coverage of the subject.

Finally, it is hoped that the student will gain an appreciation of the great beauty and perfection of normal embryonic and fetal development.

To my parents
Barbara and Robert Middleton

N.B. * A measurement in millimetres of an embryo/fetus from the crown of its head to its rump.
** Staging based on *Developmental Stages in Human Embryos* by O'Rahilly and Muller, Carnegie Institution of Washington Publication 637, 1987.

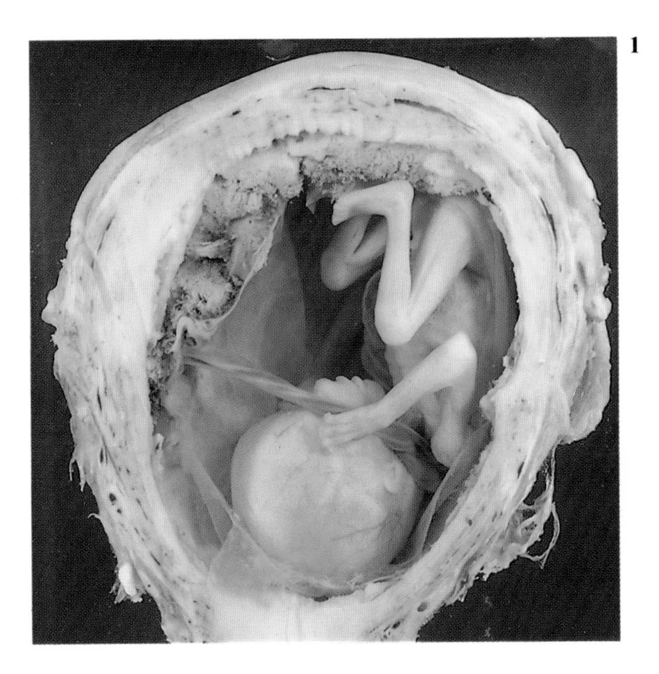

1

1 (a) Where does implantation normally occur?
(b) What is an ectopic pregnancy?
(c) Where would a placenta praevia be found?

2 (a) How many arteries and veins are present in the fetal umbilical cord?
(b) Which carries oxygenated blood?
(c) Where is the umbilical artery(s) retained in the adult?
(d) What are 'false knots'?

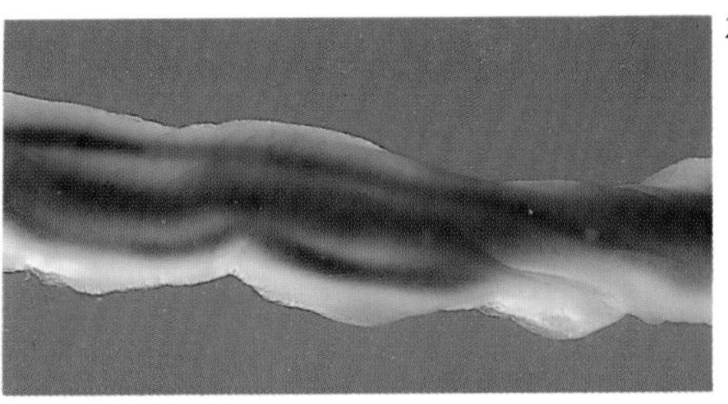

2

3

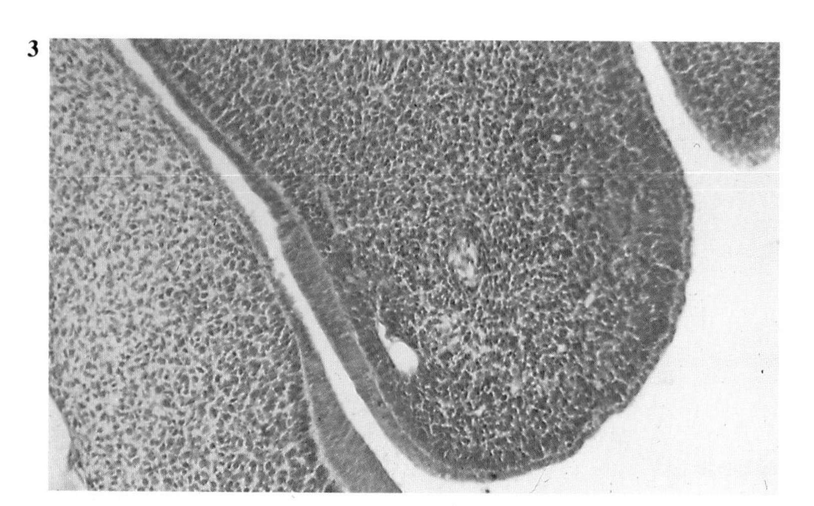

3 (a) Describe the composition of a branchial arch.
(b) What will form from mesenchyme and neural crest cells?
(c) What are the branchial arch nerves?

4 Coronal section of the cerebral hemispheres.
(a) What is the hollow vesicle in each cerebral hemisphere?
(b) How does each connect with the third ventricle?
(c) What does the original forebrain (prosencephalon) cavity form in the adult?

4

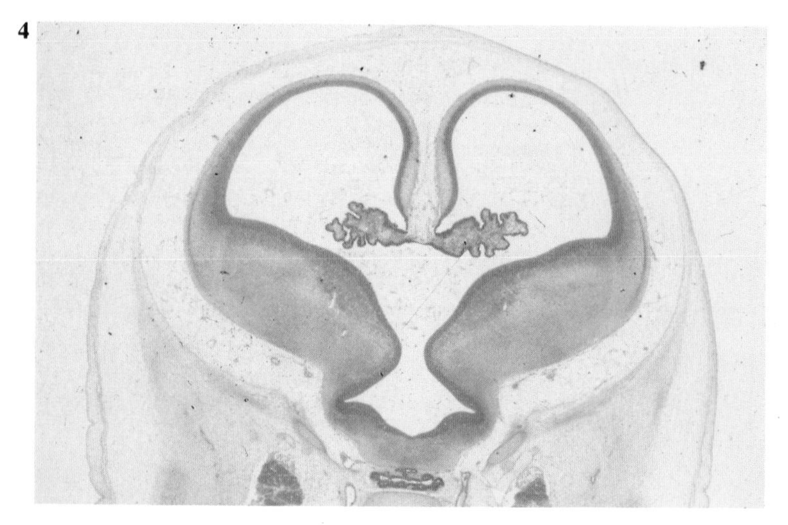

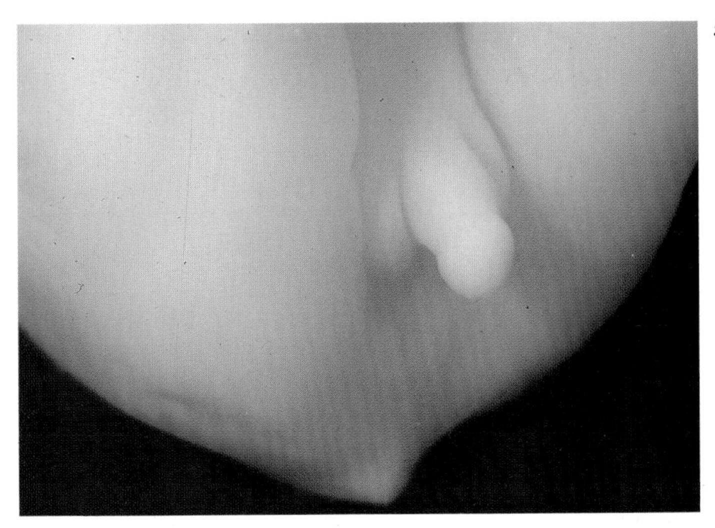

5

5 The human tail is a transient structure.
(a) From what does it develop?
(b) By what process does it disappear?
(c) At term, what structure normally represents the tail remnant?

6 (a) What is syndactyly?
(b) What are meromelia and amelia?
(c) Which limb defects are more common: major or minor?

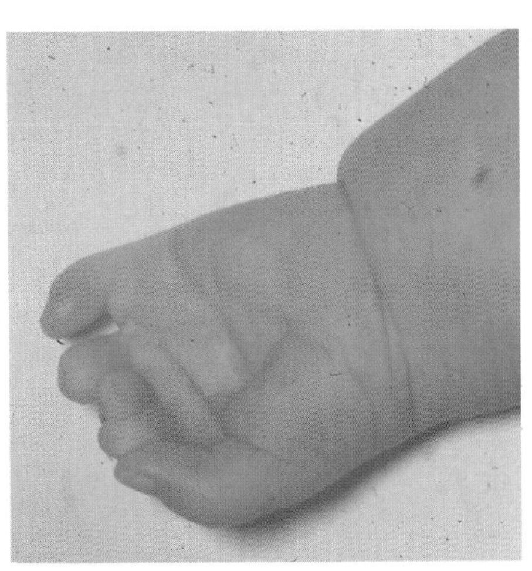

6

7

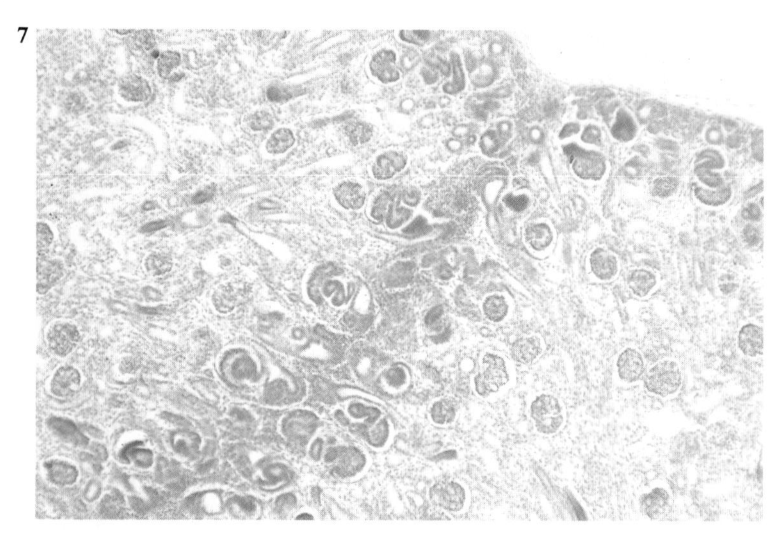

7 The kidney of a 105mm CR* fetus. (*See page 4*)
(a) What structures are present in the midline band?
(b) Is the band a normal structure?
(c) When are the neonatal glomeruli fully functional?

8 Lung buds of a 14mm CR (Stage 17)** embryo. (***See page 4*)
(a) How early do they form in the embryo?
(b) They are derivatives of which primary embryonic layer(s)?

8

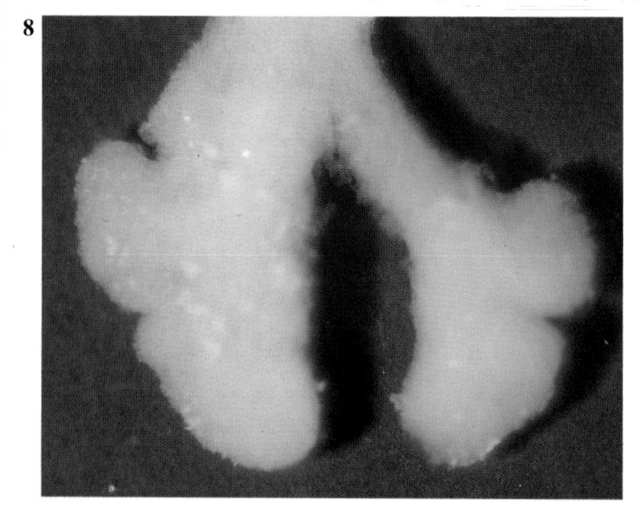

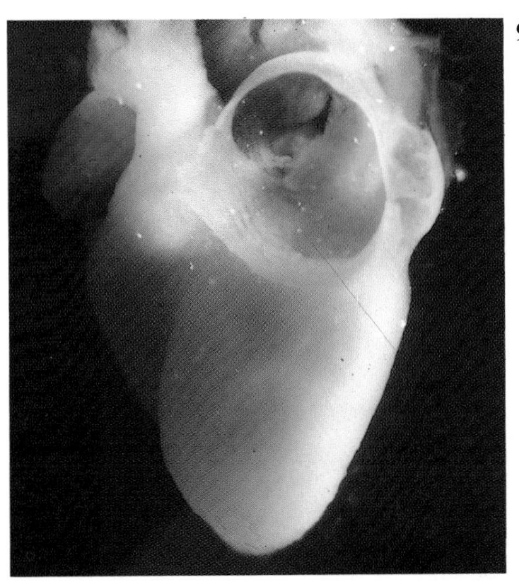

9 The heart viewed from the left side with the left atrium dissected open.
(a) From which embryonic structure are the smooth walls of the left atrium derived?
(b) Where does the blood entering the left atrium come from?
(c) At birth does the pressure increase significantly in the right or the left atrium?

10

10 (a) Where do the urogenital folds fuse in the female?
(b) What do they form in the adult female?
(c) What happens to the labioscrotal swellings in the female?

11

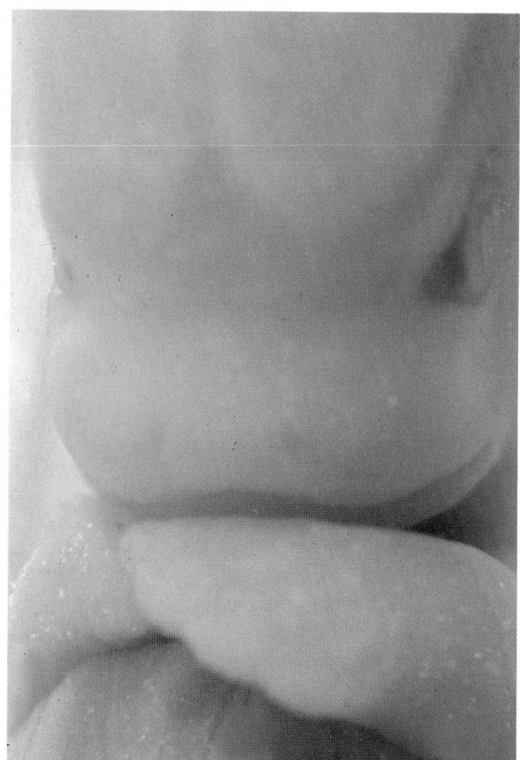

11 The face of a 20mm CR (Stage 19) embryo.
(a) How do the eyes and nostrils assume their final positions?
(b) Which embryonic structure forms the bridge of the adult nose?
(c) When does development of the face mainly occur?

12 This bone is the earliest to ossify.
(a) Which type of ossification is involved?
(b) What is the other type of ossification?
(c) In which sex do the centres of ossification generally fuse first?

12

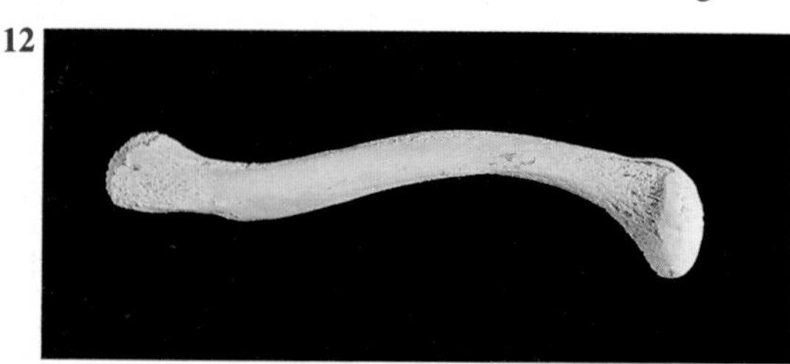

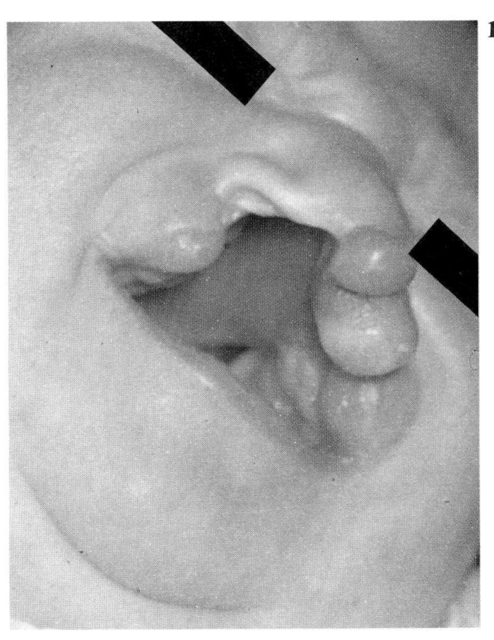

13 (a) Which normal process failed to occur here?
(b) How does the intermaxillary segment normally receive its nerve supply?
(c) If fusion is incomplete, how is the intermaxillary segment supplied by nerves?
(d) What is the frequency of cleft lip?

14 (a) Identify the diamond-shaped area in the skull vault.
(b) When does it close?
(c) How does it close?

14

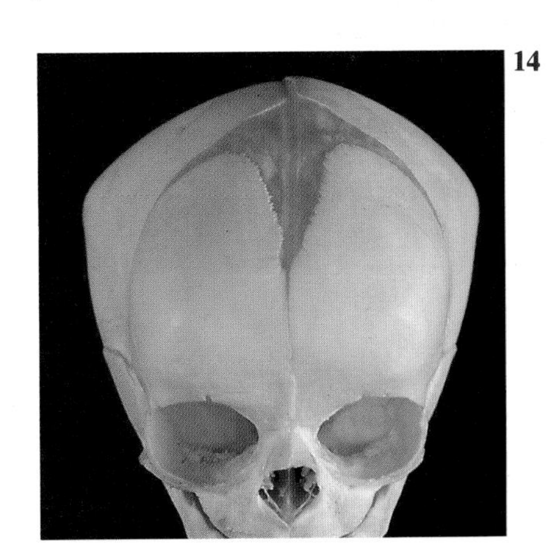

15

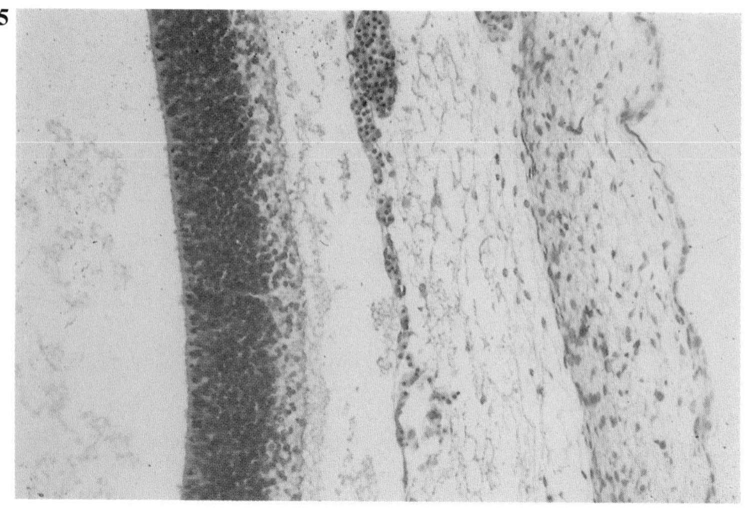

15 (a) From what structure are the membranes covering the neural tissue derived?
(b) What layer does the outer part form?
(c) Which other tissue makes a contribution to the inner layer?

16 The olfactory bulb (*arrow*) of a 57mm CR fetus viewed from the medial aspect.
(a) Which cranial nerve arises from the bulb?
(b) When do the cranial nerves form?
(c) What are the three different embryological origins of the cranial nerves?

16

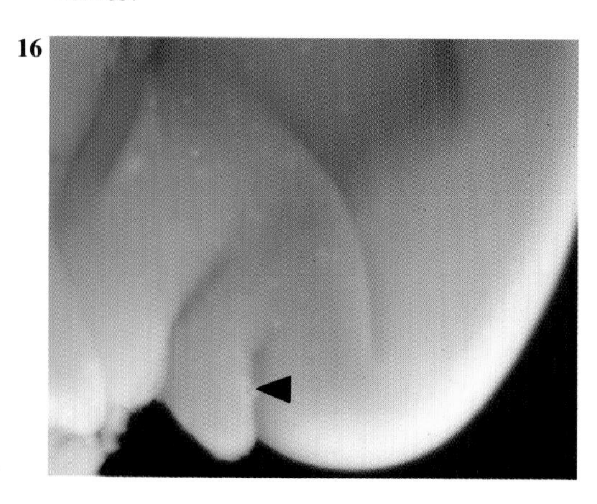

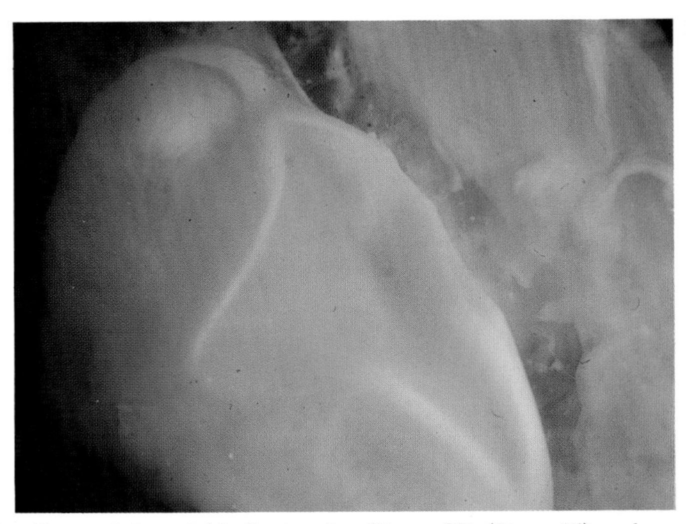

17

17 The diamond-shaped hindbrain of a 12mm CR (Stage 16) embryo viewed from above and behind.
(a) Which part of the embryonic brain contributes to the medulla oblongata?
(b) The alar and basal laminae can be distinguished clearly in histological sections. By which landmark are they separated?
(c) How do the pyramids form?

18 (a) What is the name of the second type of nephric tubules which form late in Week 4?
(b) From which tissue does the mesonephric duct form?
(c) Into what structure does the mesonephric duct open?

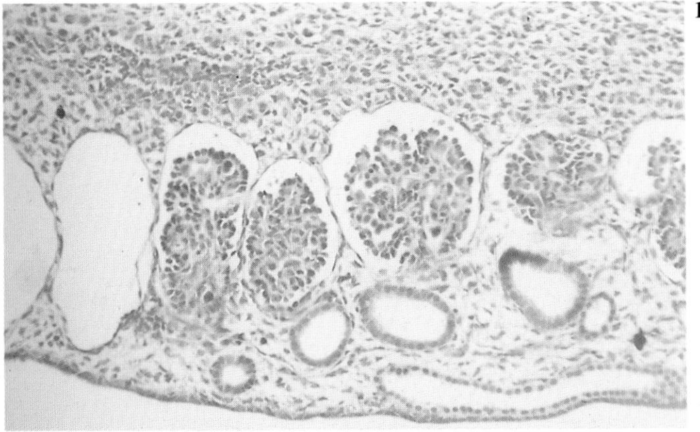

18

19

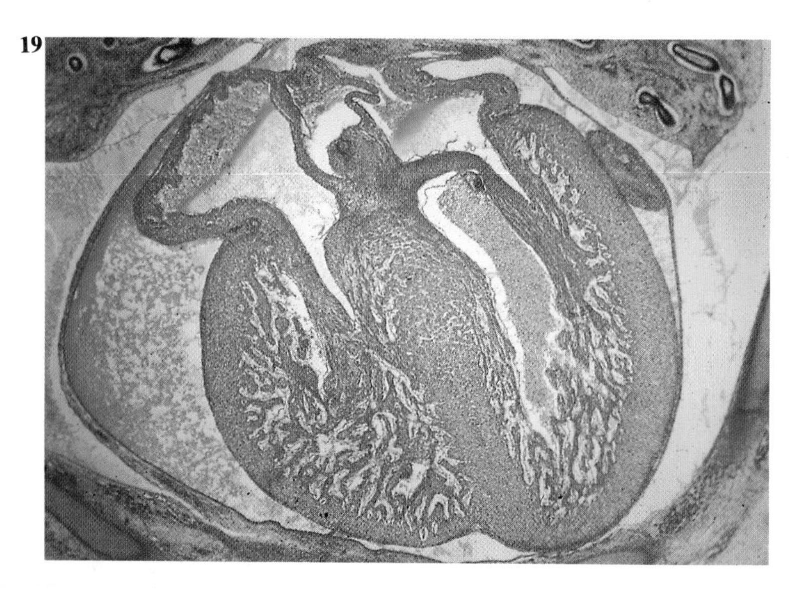

19 Transverse section of the heart.
(a) Which embryonic structure gives rise to the ventricles?
(b) Which primary embryonic tissue gives rise to the heart?
(c) Name the tissues which contribute to the interventricular septum.

20 Transverse section of skin and developing mammary gland from a 155mm CR fetus.
(a) How does this developing gland form?
(b) How does the nipple form?

20

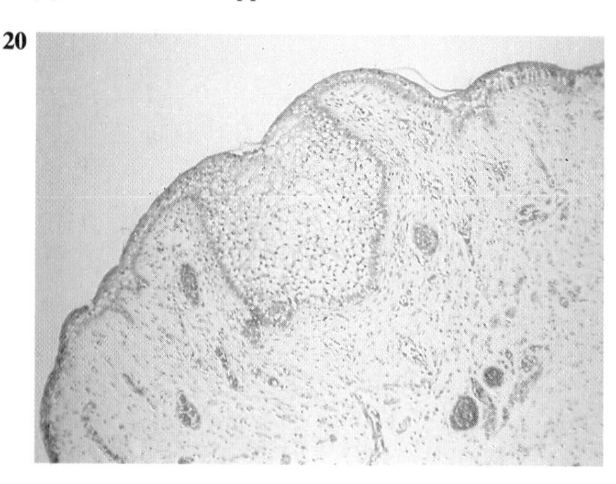

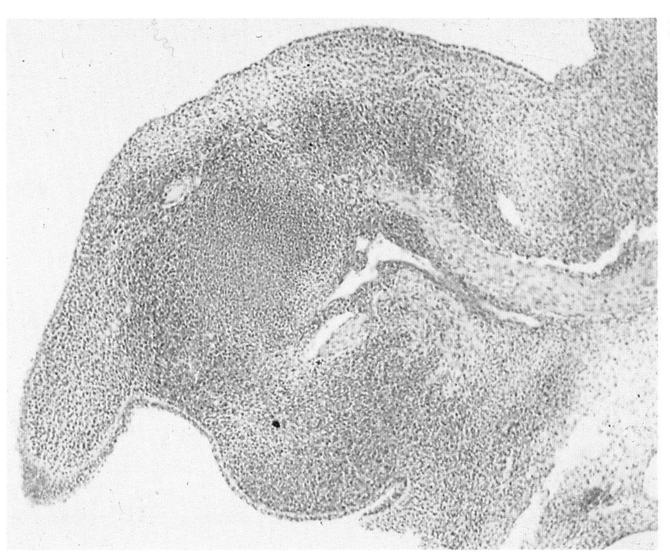

21

21 Coronal section of the arm bud.
(a) Which nerves supply the flexor and extensor compartments of the arm bud?
(b) What do the preaxial and postaxial borders become in the adult?
(c) How do the somatic muscles of the limb develop?

22 (a) What is the origin of the muscle of the diaphragm?
(b) Name the common congenital defect arising from failure of the formation or fusion of the pleuroperitoneal membranes.
(c) What anatomical disorder is often associated with this defect?

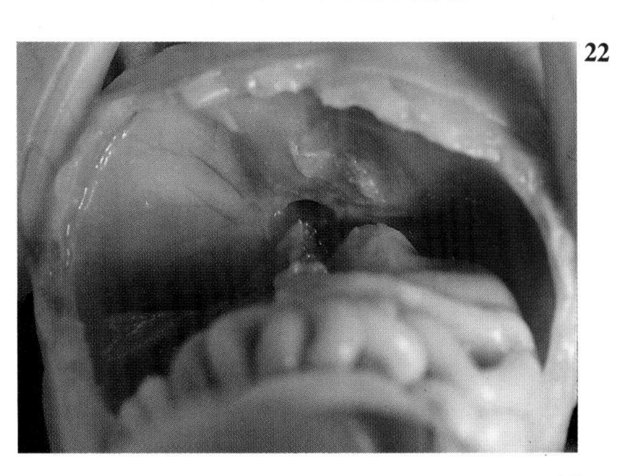

22

23

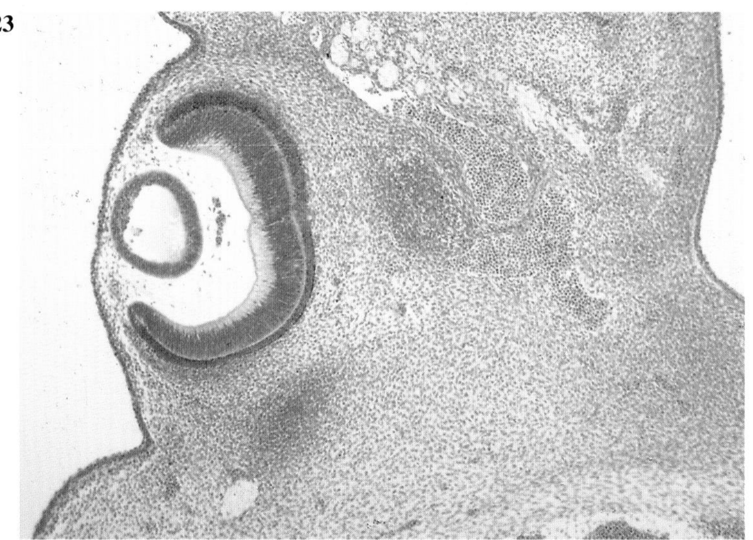

23 (a) Which structure gives rise to the optic vesicles?
(b) Why are the optic and olfactory cranial nerves atypical in origin?
(c) What does the optic vesicle form in the adult?
(d) When does myelination of the optic nerves occur?

24 (a) At birth how well developed is the musculature of the anal sphincter?
(b) What is the dual origin of the anal canal?

24

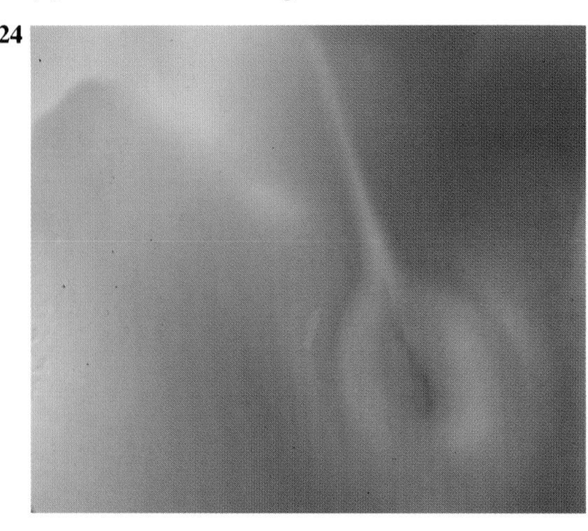

25

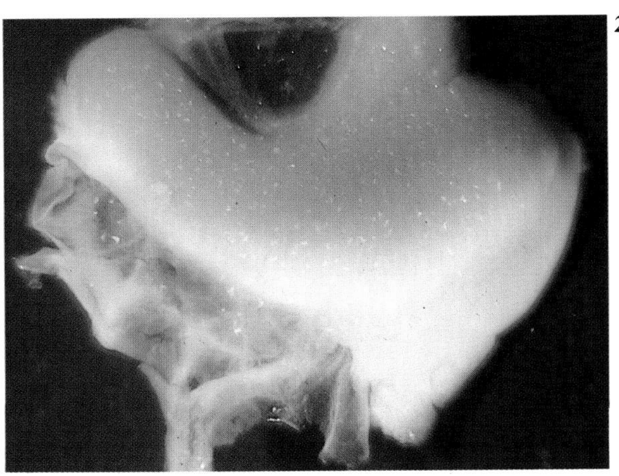

25 The stomach, spleen and greater omentum of a 123mm CR fetus.
(a) Where does the spleen develop?
(b) What three types of cells are present in the spleen when haemopoiesis reaches a peak in Week 13-20?
(c) Accessory spleens are common anomalies. Where do they usually occur?
(d) Which other fetal organ affects the development of the white pulp of the spleen?

26 Longitudinal view in an ultrasound scan.
(a) Identify this anomaly.
(b) This anomaly is confirmed by finding increased levels of a certain protein in the amniotic fluid. What is the name of the protein?
(c) Can this fetus empty its bladder and kick *in utero*?

26

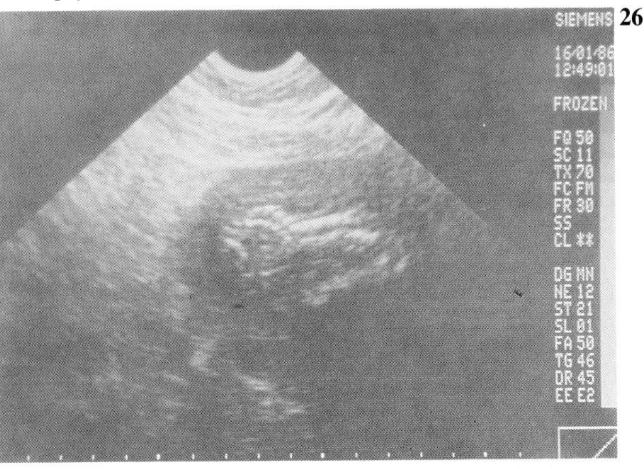

27

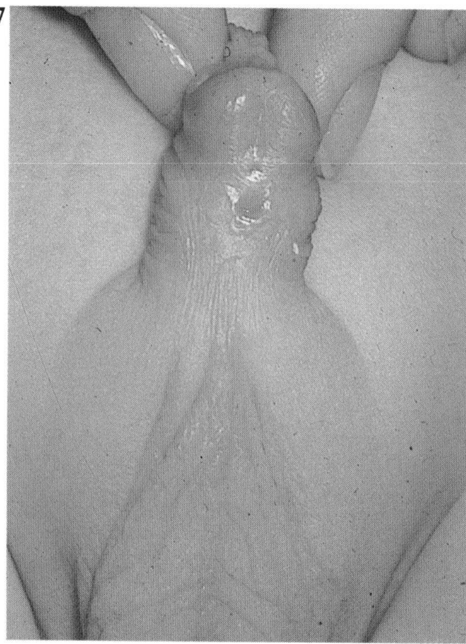

27 (a) What is epispadias in the male?
(b) What is hypospadias in the male?
(c) Which hormone deficiency is responsible for hypospadias?

28 The caecum and appendix of a 60mm CR fetus.
(a) The caecal diverticulum marks the border between which two parts of the gut?
(b) How does the appendix form?
(c) How do the caecum and appendix assume an adult position?
(d) Is there a sharp division between the caecum and appendix in the neonate?

28

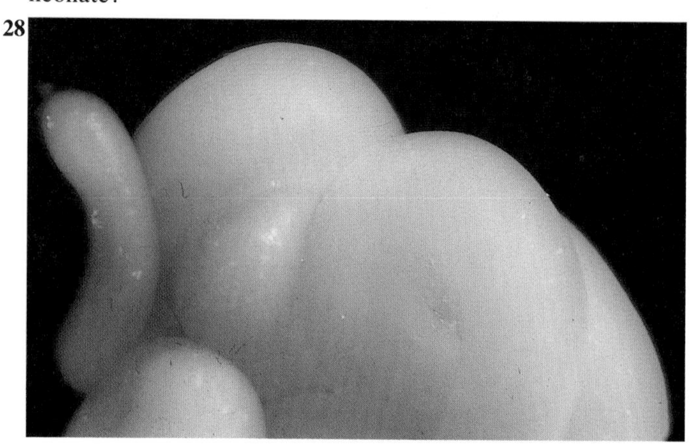

29

29 The larynx viewed from behind.
(a) Which embryonic tissues contribute to the larynx?
(b) What are the three swellings surrounding the opening to the trachea?
(c) Which branchial arches form the laryngeal muscles and cartilages?

30

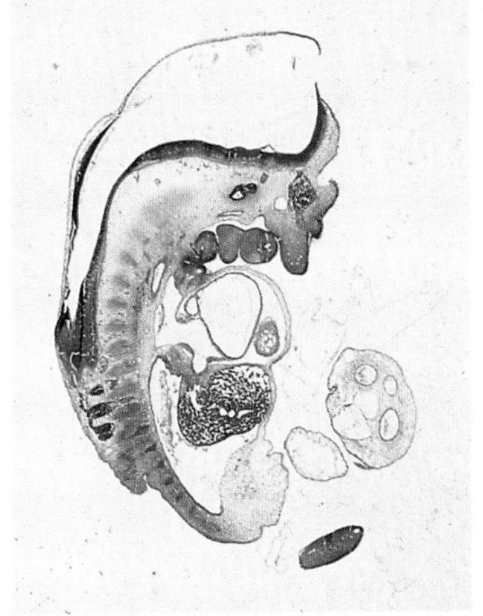

30 A sagittal section through an early embryo.
(a) What is the venous sinus from which blood enters the primitive atrium?
(b) Which vessels drain into this sinus?
(c) Where are the arterial and venous ends of the heart fixed in position?

31

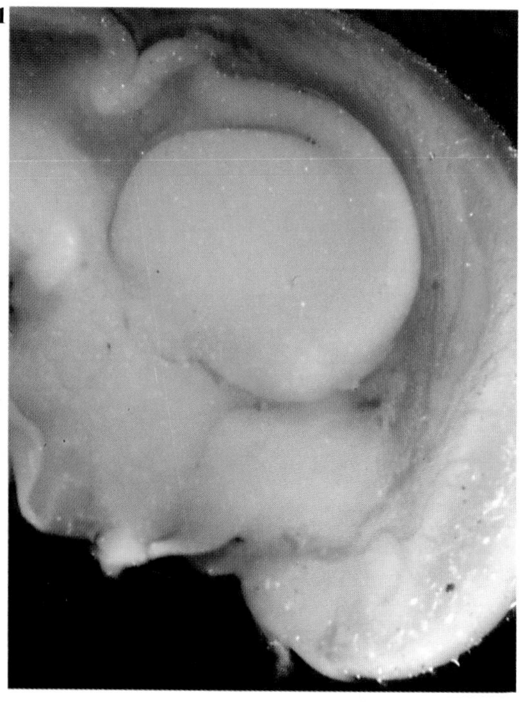

31 Sagittal section of the brain of a 57mm CR fetus viewed from the medial aspect.

(a) What three swellings appear in the walls of the third ventricle?

(b) Of what tissue is the epithalamus a derivative?

(c) What is the massa intermedia?

32

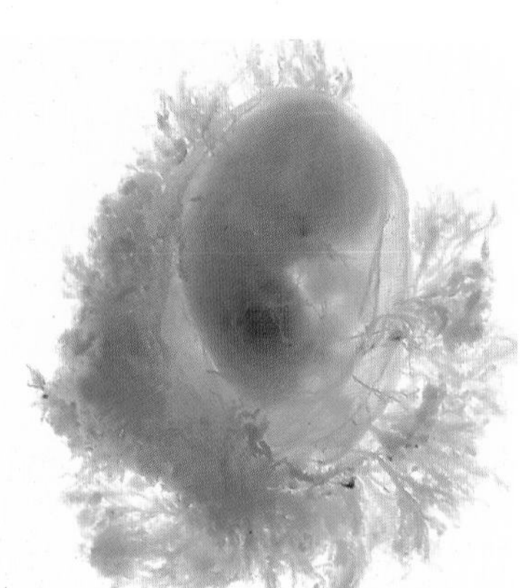

32 Up to Week 8 the chorionic sac is covered with villi.

(a) What is the immediate fate of the decidua capsularis?

(b) Which embryonic membrane lines the chorion?

33

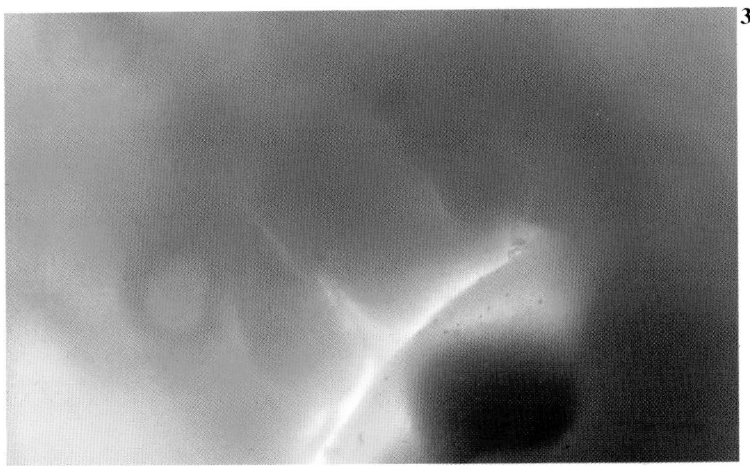

33 Side view of the branchial region of a 12mm CR (Stage 16) embryo.
(a) What are the alternative names for the first and second branchial arches?
(b) What are the prominences of the first arch?
(c) To what structures does the second arch contribute?

34 An ultrasound scan of a fetus at an estimated gestational age of 30.5 weeks, based on a femoral length of 59mm.
(a) From what part of the femur are measurements made to determine the length of this bone?
(b) When can the femoral length be measured accurately?
(c) What other measurements correlate with the femoral length?

34

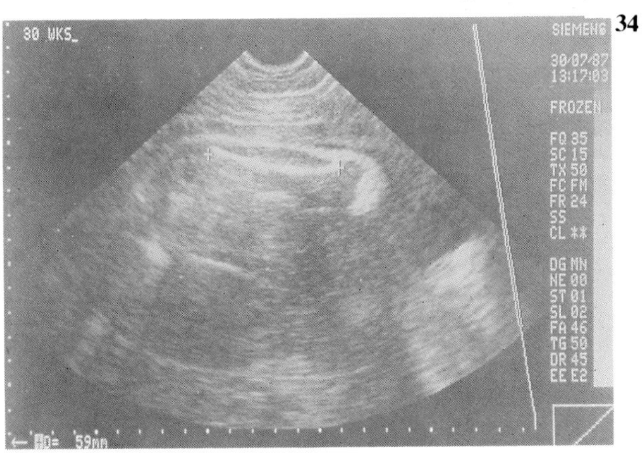

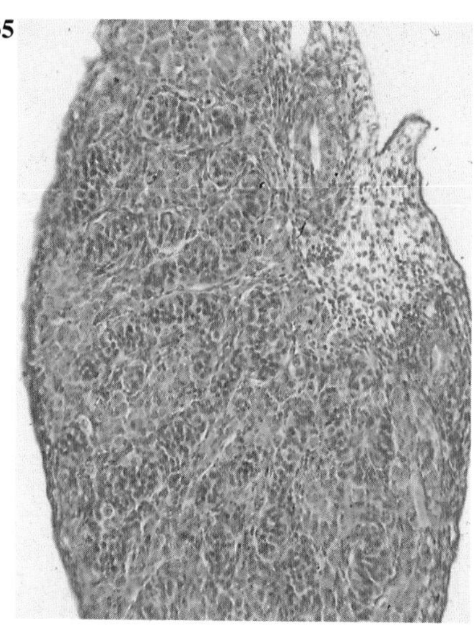

35 The developing gonad in a 70mm CR fetus.

(a) Is this an ovary or a testis?

(b) After what age do the ovaries and testes differ in appearance?

(c) Coelomic epithelial cords (primary sex cords) grow into the mesoderm of the gonadal region. Which parts are formed by them?

(d) What is the ultimate fate of the cords in both sexes?

36 (a) What is a likely cause of the honeycomb appearance on this X-ray?

(b) How is it thought to form?

(c) Which other organs may be similarly affected?

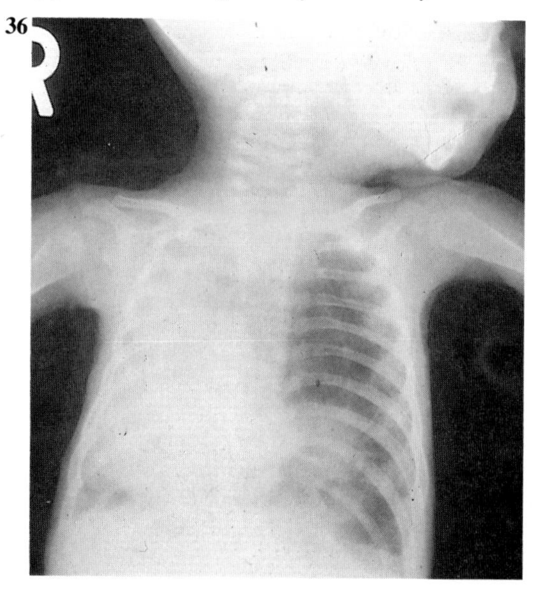

37 The ventral pancreatic bud is carried medially during gut rotation where it fuses with the dorsal pancreatic bud. The fused buds (*arrow*) are shown below a red-coloured needle.

(a) What are the derivatives of the ventral bud?

(b) What are the derivatives of the dorsal bud?

(c) How do the pancreatic ducts form?

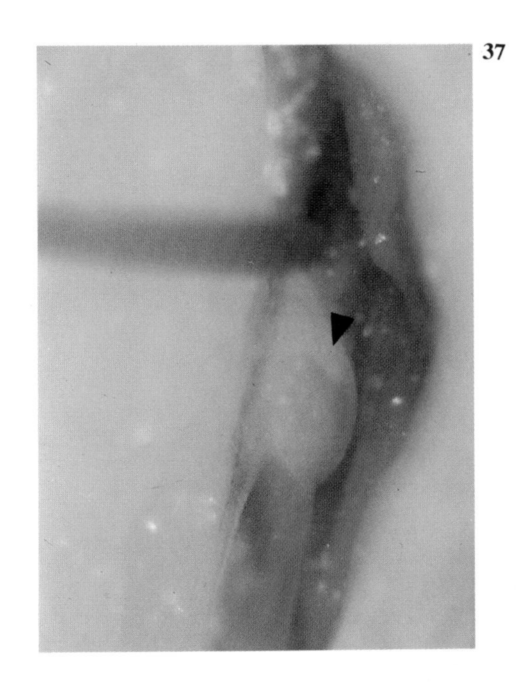

37

38 The hypophysis (*arrow*) in the sella turcica.

(a) Name the dual origin of this tissue.

(b) What develops from the adenohypophysis?

(c) What develops from the neurohypophysis and its cavity?

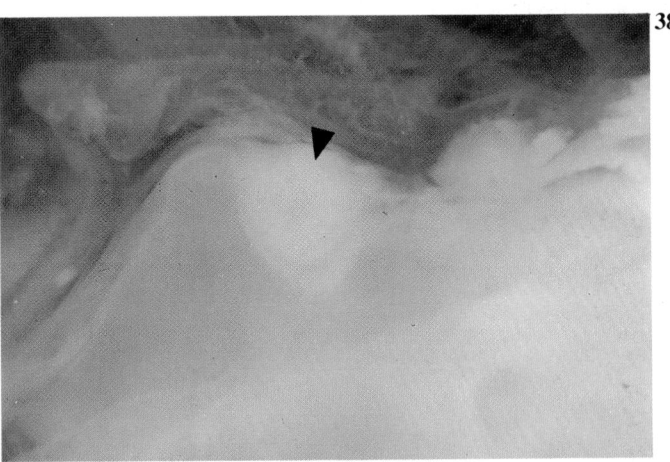

38

39

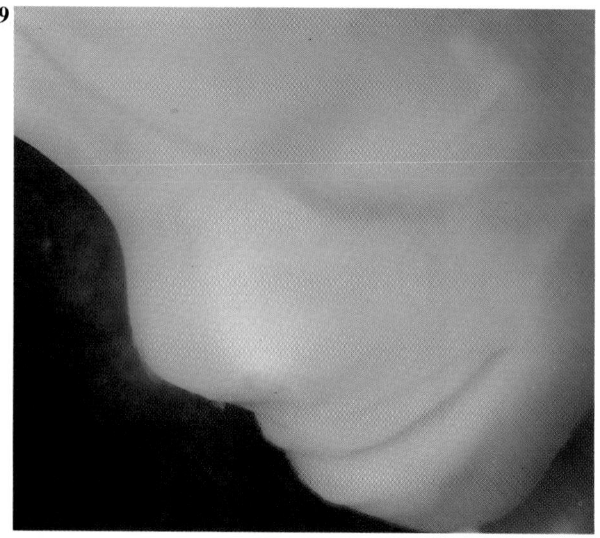

39 The face of a 51mm CR fetus.
(a) How is the broad mouth opening reduced in size?
(b) Which branchial arch forms the muscles of facial expression?
(c) Which branchial arch forms the muscles of mastication?
(d) Can chronic maternal use of alcohol have an effect upon the development of the face?

40 (a) What fluid do these plexuses produce?
(b) Where are they located in the brain?
(c) How do they form in the ventricles?

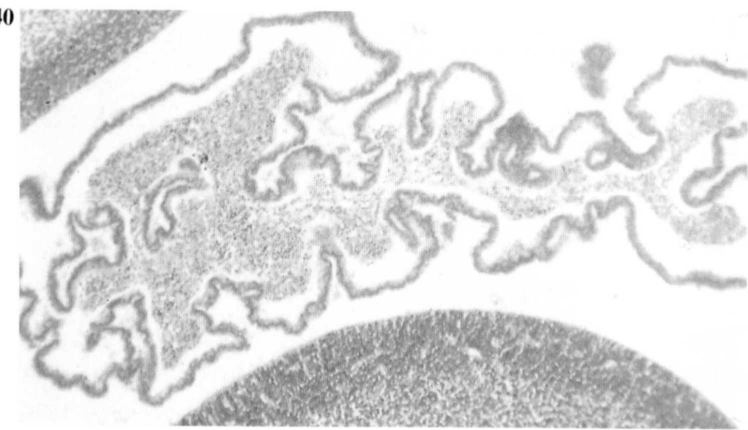

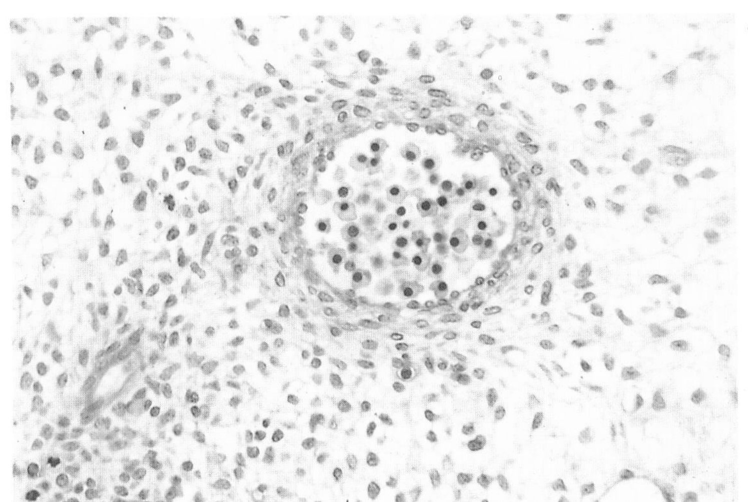

41

41 In the early embryo the red blood cells seen in the blood vessel are normally nucleated.
(a) When do fetal red blood cells normally become anucleate?
(b) When are the blood group antigens expressed in the embryo?
(c) What tissue will be formed from the mesenchyme cells condensing around the vascular endothelium?

42 (a) What are the constituents of the embryonic stalk?
(b) What is the connective tissue derived from mesenchyme called?
(c) What major event transforms the embryonic stalk into an umbilical cord?

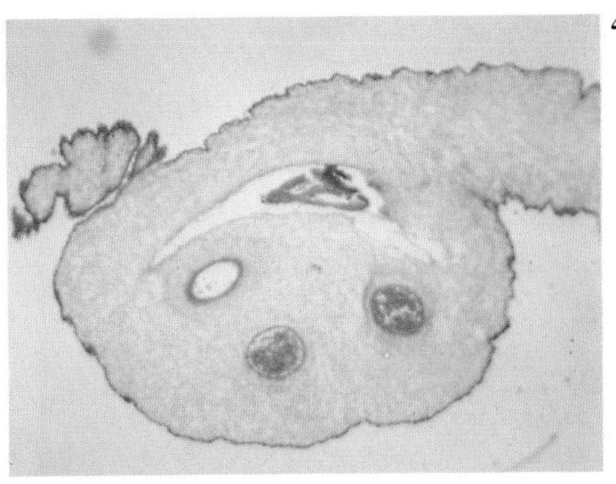

42

43

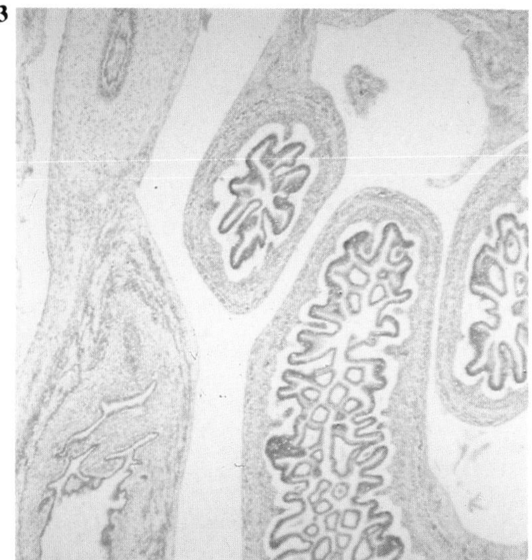

43 The bladder and allantois (on the left) of a 40mm CR fetus.
(a) As the bladder forms, what is the fate of the allantois?
(b) When the bladder enlarges, what two structures are incorporated into the dorsal wall and enter the bladder separately?
(c) Is the neonatal bladder pelvic or abdominal in position?

44 (a) Does the mandible or maxilla fuse first?
(b) How many bones fuse to form the mandible in the adult?
(c) What is the embryonic structure which serves as scaffolding for the mandible?
(d) What is the approximate angle of the rami in the neonate?

44

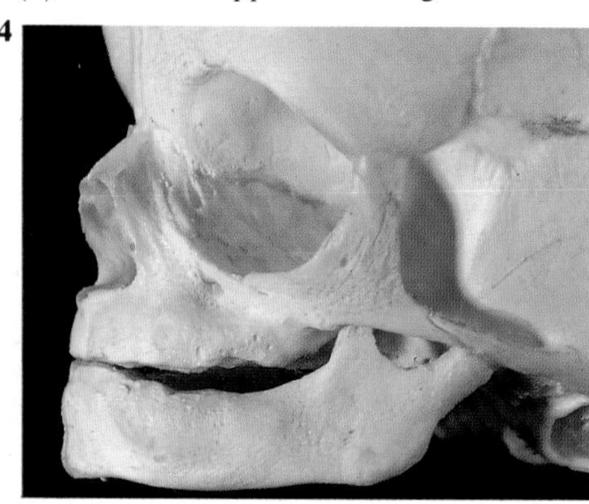

45 The early oesophagus and stomach.
(a) From what tissue does the oesophagus derive?
(b) Which tissue will form muscle and the submucosal layers?
(c) How does it elongate during development?

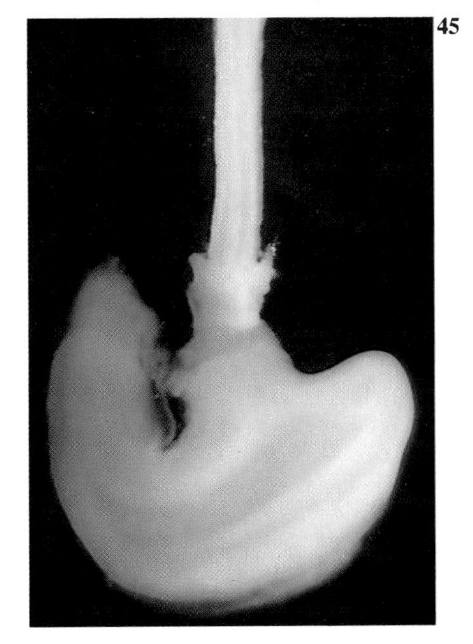
45

46 Identify the various parts of the arm.
(a) What are the future digits called at this stage?
(b) What are the areas between the future digits called?
(c) What is the term for the entire future hand?

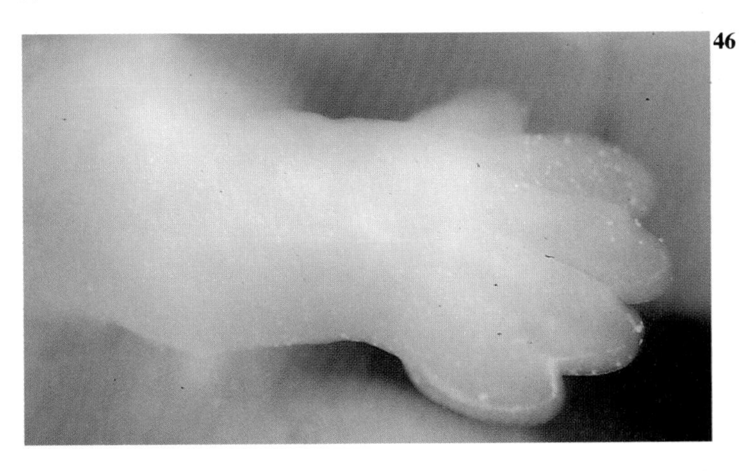
46

47

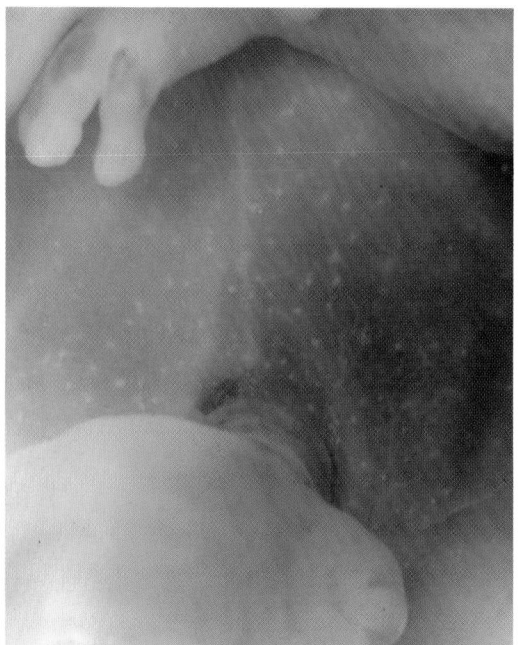

47 Midgut herniation in a 27mm CR (Stage 22) embryo.
(a) Which limb of the midgut herniation first returns to the abdominal cavity?
(b) During midgut rotation what normally happens to the yolk stalk?
(c) Which embryonic structure persists as Meckel's diverticulum?

48 Adult interatrial septum viewed from the right atrium.
(a) The floor of the fossa ovalis is derived from which embryonic structure?
(b) What is the opening present between the two atria at birth?
(c) From which embryonic structure is the limbus fossae ovalis derived?

48

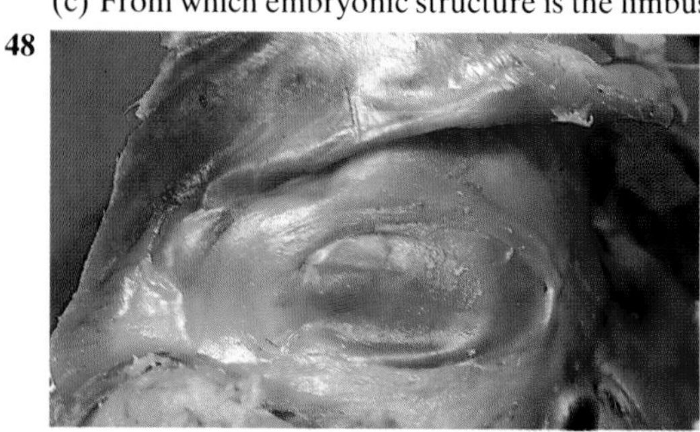

49 Dorsum of the tongue and lower lip.
(a) The tongue has two major contributions to its development. What are they?
(b) What features separate the anterior two-thirds from the posterior third of the tongue?
(c) What is the fate of the tuberculum impar?

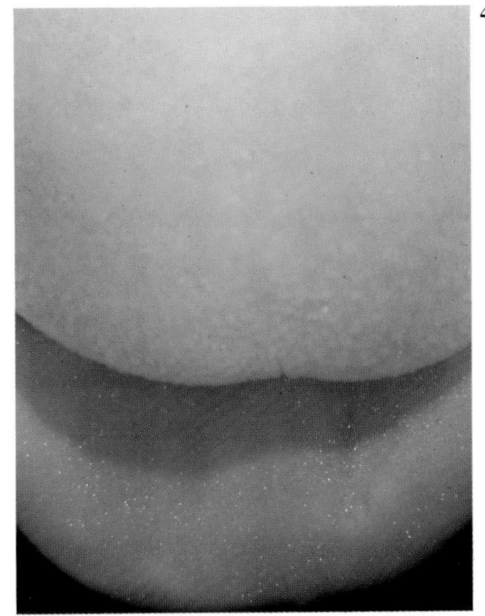

49

50 (a) Identify this congenital anomaly.
(b) What has failed to occur during development?
(c) Is this a commonly occurring abnormality?

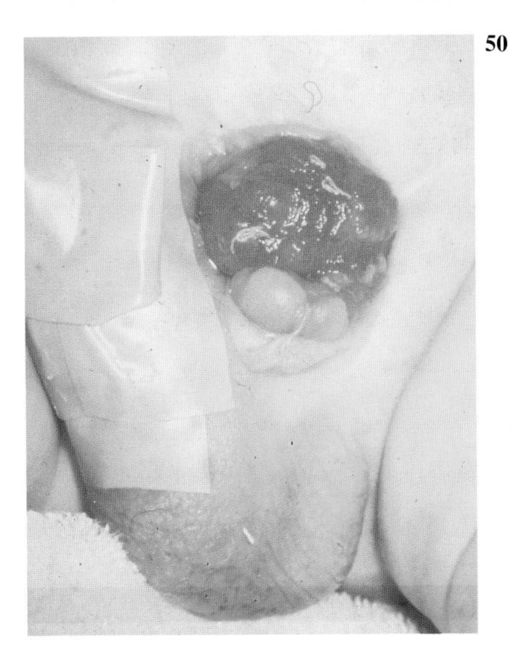

50

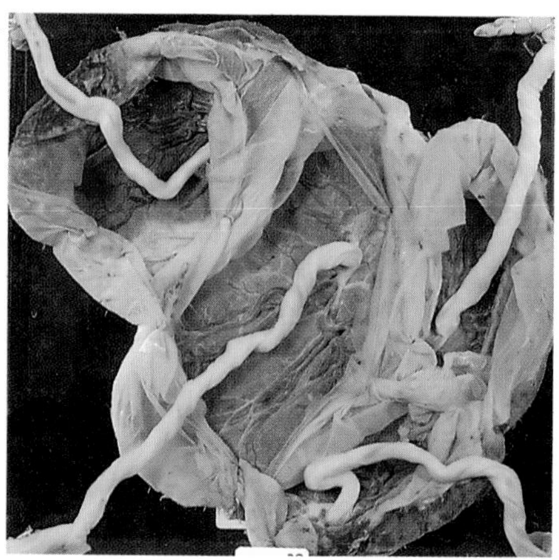

51 Placentas from quadruplets.

(a) What are multiple births derived from separate zygotes called?

(b) What are multiple births derived from one zygote called?

(c) In multiple pregnancies, how are the fetal membranes and placenta(e) related to the derivation of the embryos?

52 The heart and its accompanying sac.

(a) Which embryonic membrane gives rise to the fibrous pericardium surrounding the heart?

(b) Name another derivative of this embryonic membrane.

(c) Which major blood vessel is initially incorporated into the membrane?

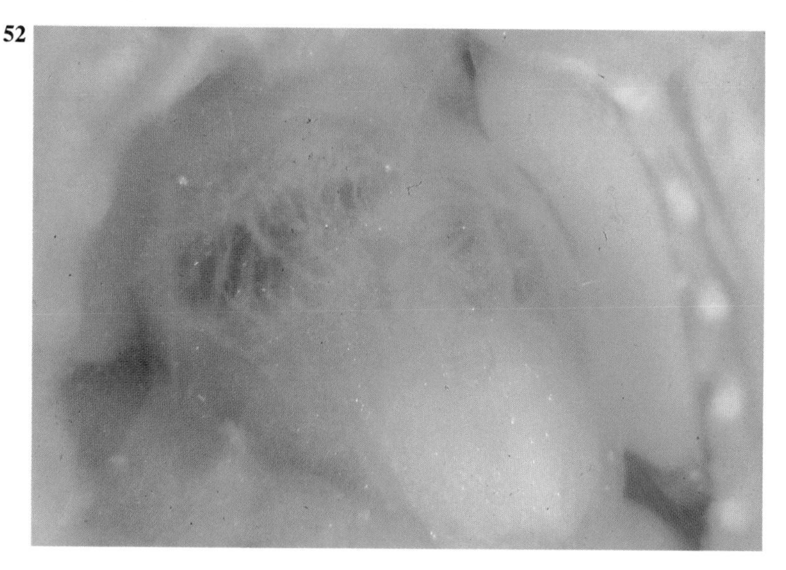

53 At Week 8 the spinal cord is present throughout the length of the embryo.
(a) How does it come to lie at a higher vertebral level as the embryo grows?
(b) Are the spinal nerves affected by the changing level of the spinal cord?
(c) When does myelination occur?

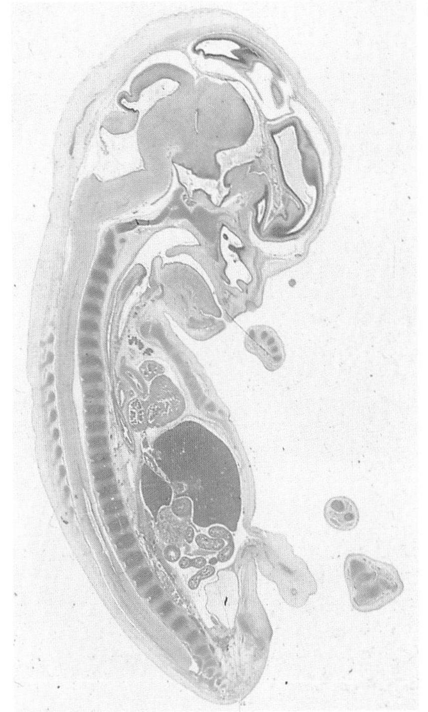

53

54 The adult female bladder and uterus.
(a) Which part of the urogenital sinus forms the bladder?
(b) What contributes to the trigone of the bladder?
(c) In the infant the urachus becomes fibrous. What does it form?

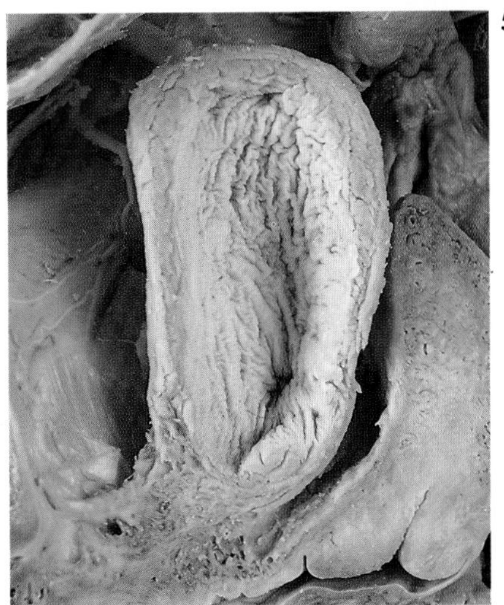

54

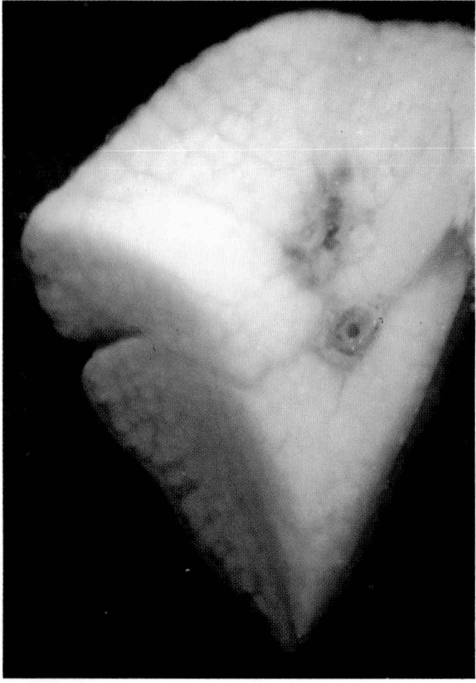

55 (a) Which features of the lung must be overcome by the first breath at birth?
(b) Which alveoli dilate first?
(c) Is a deficiency of surfactant associated with any disease?
(d) State which hormone stimulates surfactant production.

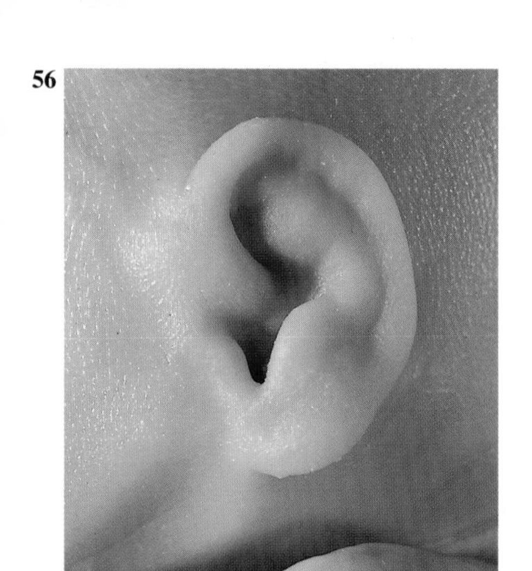

56 (a) Each auricle of the ear is formed from six tubercles. Which branchial arches are involved?
(b) Which adult structure is derived from the most ventral tubercle of the first arch?

57 (a) The spinal cord forms by two processes. Which one occurs at the caudal end of the cord?
(b) Which tissue separates from the margin between the forming neural tube and adjacent ectoderm?
(c) Name the fixed curves of the vertebral column at birth.

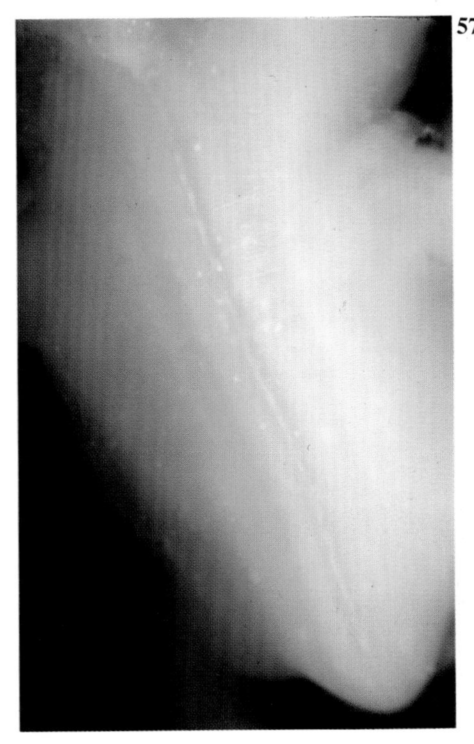
57

58 The spinal cord and ganglia of a 56mm CR fetus.
(a) What are the primordia of the ventral roots of the spinal nerves?
(b) What are the primordia of the dorsal root ganglia?

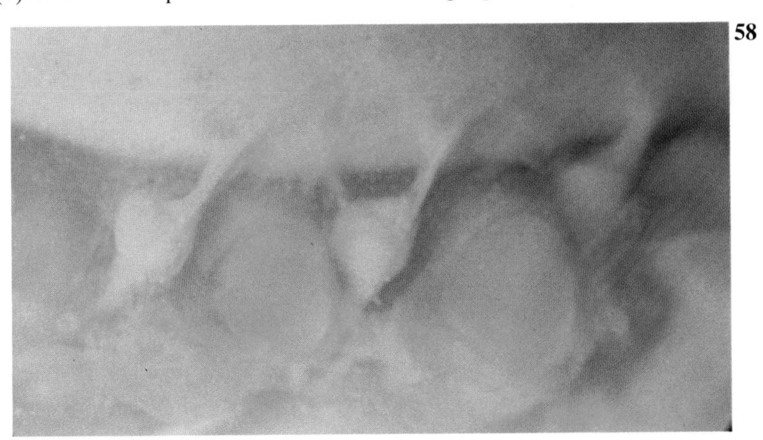
58

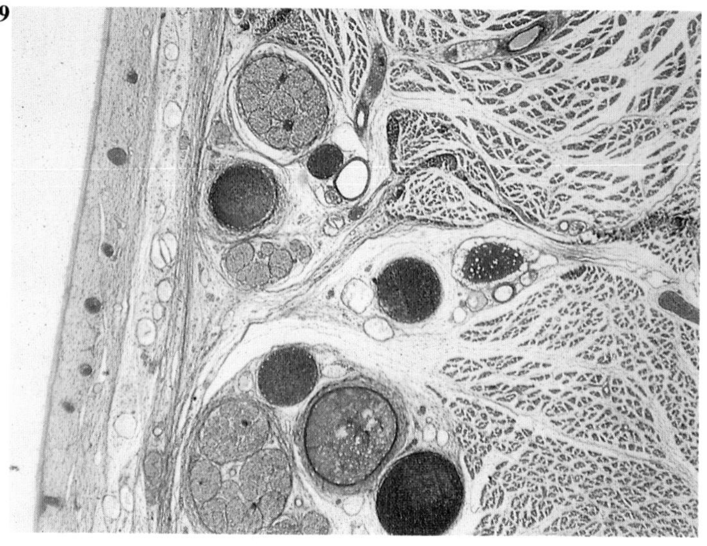

59 Limb muscles of a 3.5 month fetus.
(a) From what do limb muscles (excluding the girdle muscles) form?
(b) Which type of mesenchyme gives rise to most smooth muscles and cardiac muscles?
(c) What is the nerve supply to the dorsal and ventral parts of each myotome as they migrate to their final positions?

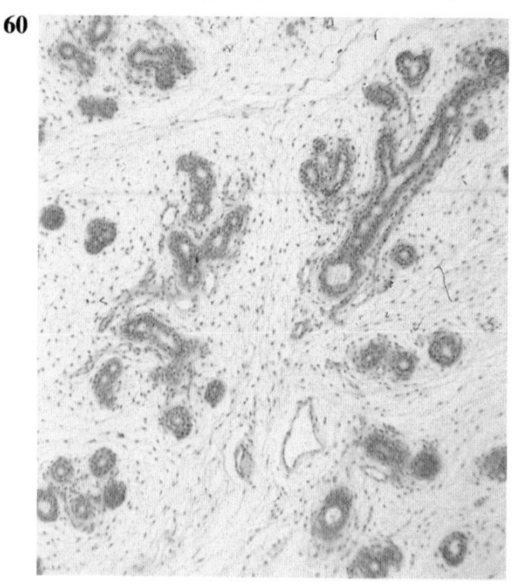

60 All three salivary glands develop from solid epithelial outgrowths into adjacent mesenchyme.
(a) Does the sublingual gland shown develop from ectoderm or endoderm?
(b) What are the derivatives of the epithelial outgrowths and the surrounding mesenchyme?
(c) In what order do the glands develop?

61

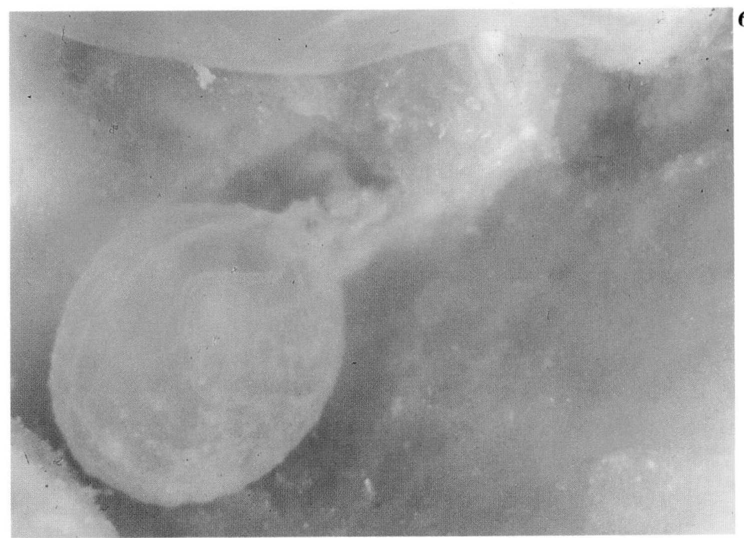

61 In Week 4 the body folds and constricts the yolk sac and a portion is incorporated into the embryo.
(a) What will the incorporated part form?
(b) What is the extra-embryonic portion called?
(c) Describe the ultimate fate of the extra-embryonic portion.

62 Face of a fetus.
(a) How do the eyelids form?
(b) The mesenchymal fold gives rise to which adult structures?
(c) The ectodermal covering gives rise to which adult structures?
(d) When do the eyelids fuse?

62

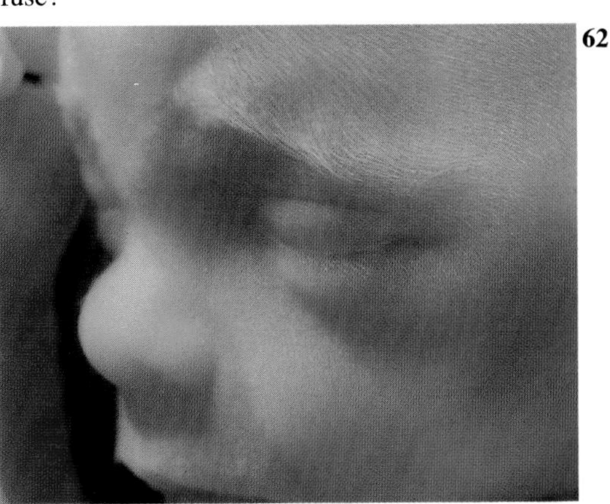

63

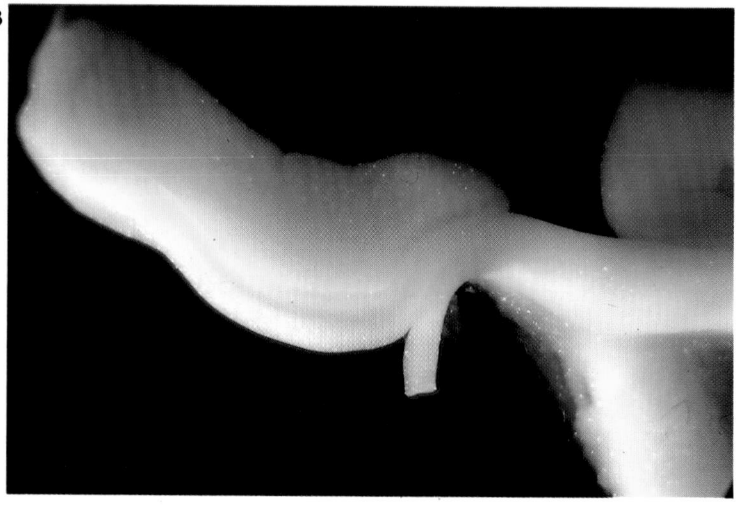

63 Ovary and uterine tube in a four month fetus.
(a) Which structure forms the ovarian ligament?
(b) From what structure does the uterine tube form?
(c) Describe the descent of the ovary.

64 (a) List the pancreatic derivatives of the endoderm.
(b) What are the derivatives of the surrounding mesenchyme?
(c) What other glands are similar in their development?

64

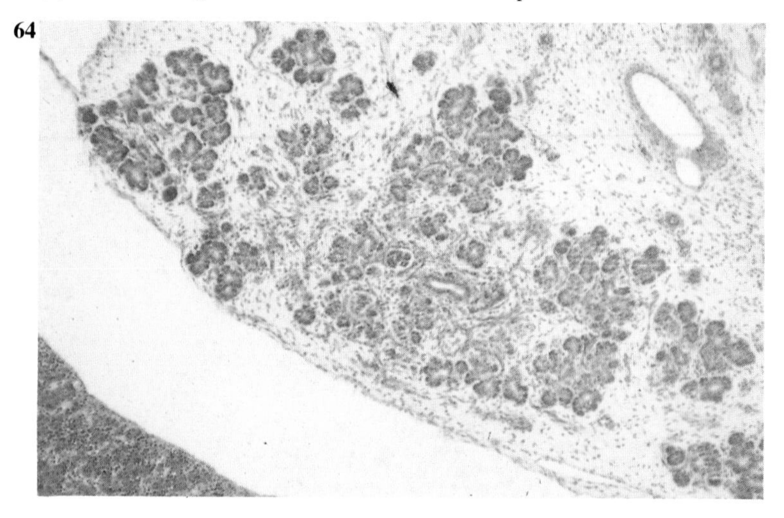

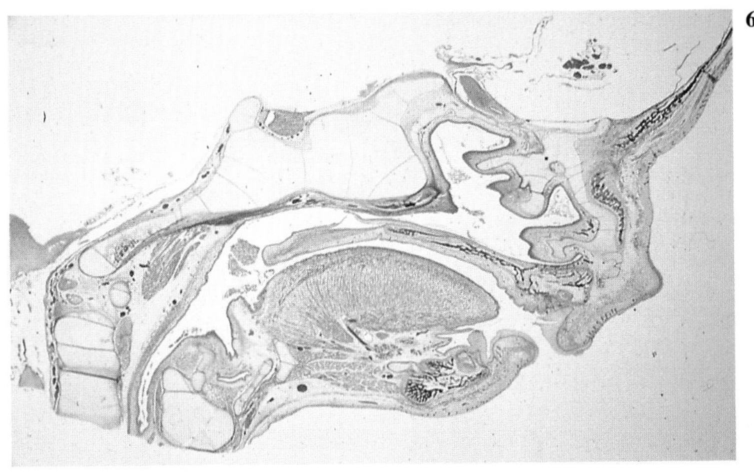

65 A sagittal section through the mouth and nose of a fetus.
(a) How does the nasal septum form?
(b) Is the olfactory epithelium ectodermal or endodermal in origin?
(c) How does the neonate breathe?

66 A hole-in-the-heart.
(a) How common are defects of the cardiac septa?
(b) Are interventricular septal defects more common in the muscular or membranous portions?
(c) What is the critical period for heart development?
(d) Describe the Tetralogy of Fallot.

66

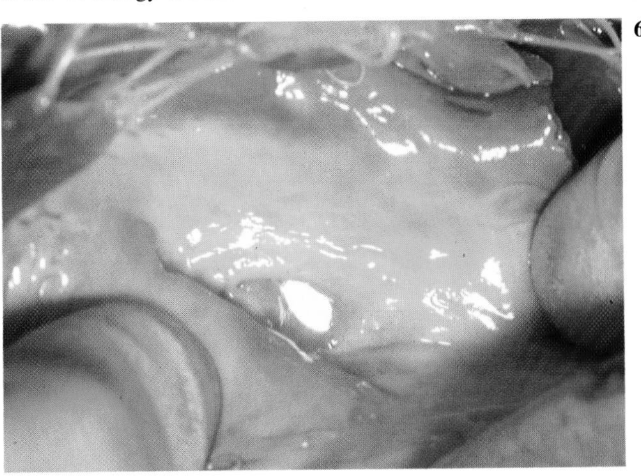

67

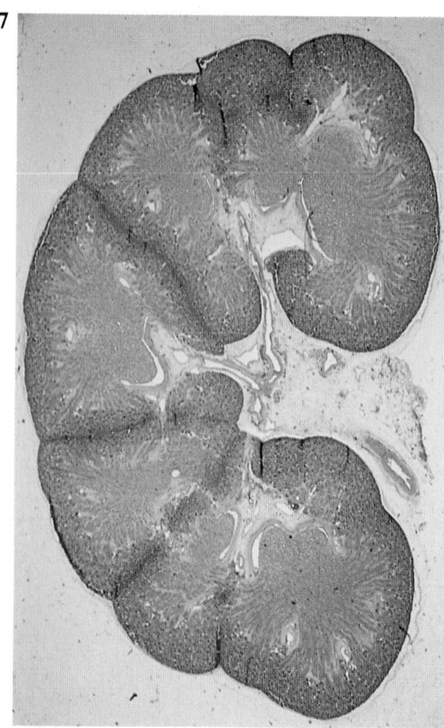

67 (a) Which way does the renal pelvis face during its apparent migration to its final position on the posterior abdominal wall?
(b) When is urine first formed?
(c) As the kidney appears to migrate cephalically, what happens to its blood supply?

68

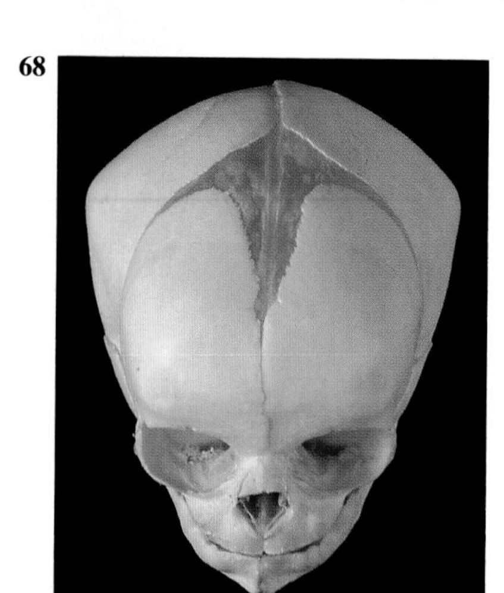

68 (a) When do the frontal bones fuse?
(b) When do the frontal sinuses appear?
(c) In early life, what is the medical use of the anterior fontanelle?

69

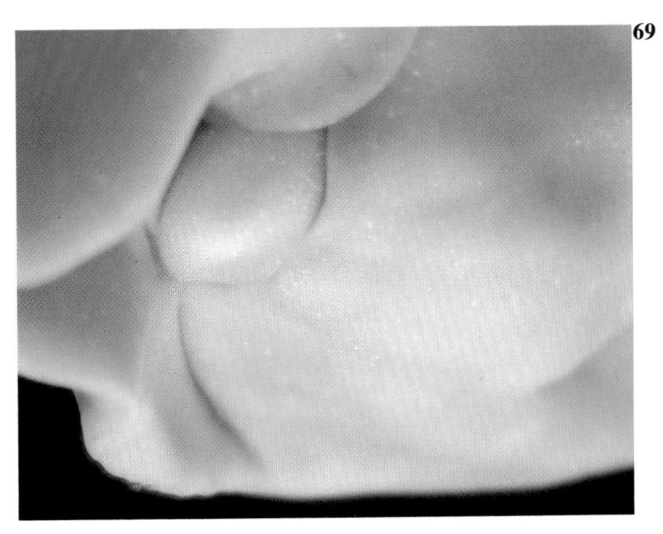

69 Dorsal surface of the fetal liver.
(a) How does the gallbladder arise?
(b) The gall·bladder stalk forms which adult structure?
(c) When does bile appear in the fetus?

70

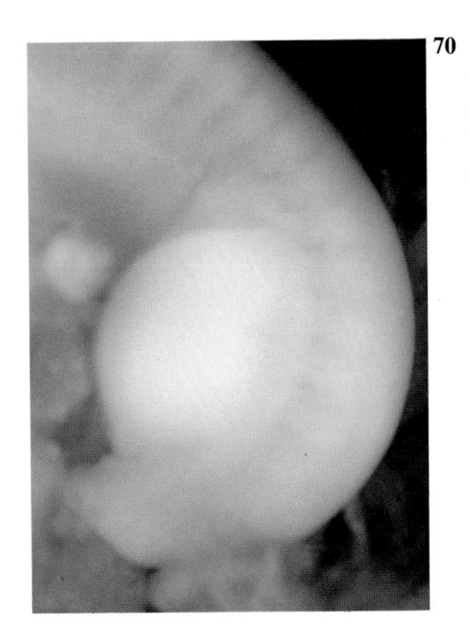

70 The back and tail of an embryo.
(a) What are the blocks of tissue?
(b) Name some of their derivatives.
(c) Head and neck muscles are exceptional in their origin. From what tissue do they arise?

71

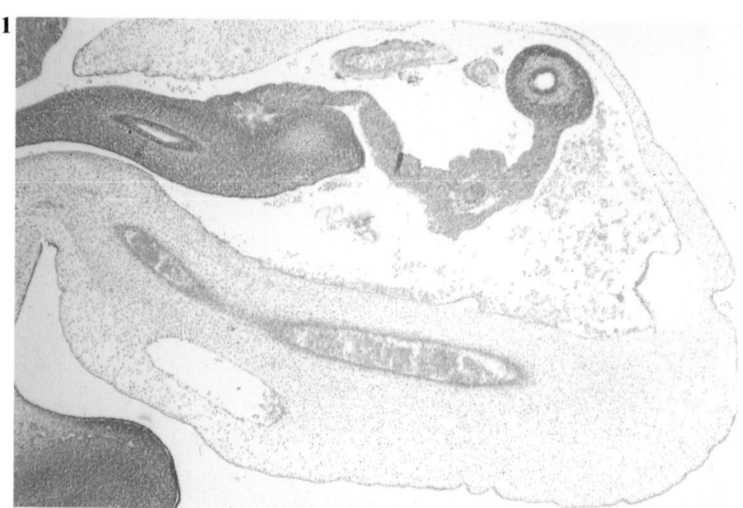

71 Umbilical stalk of a 20mm CR (Stage 19) embryo.
(a) When does the midgut herniation return to the abdominal cavity?
(b) Which adult structures are formed from the proximal and distal limbs?
(c) Which mesenteries are retained in the adult?

72 Fingernails of a 45mm CR fetus.
(a) Where do the nail fields originally develop?
(b) The nail is initially covered by a layer of epidermis. What is it called?
(c) What is the remnant in the adult called?

72

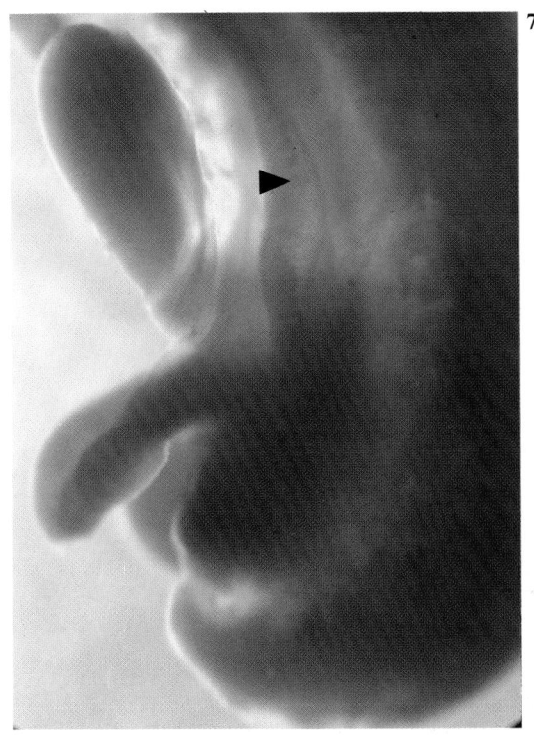

73 A 14mm CR (Stage 16) embryo viewed from the left side. The body wall has been dissected away and the mesonephric tubules (*arrow*) exposed to view.
(a) What are the derivatives of the mesonephric tubules in the male?
(b) What are they in the female?
(c) What is the fate of the mesonephric duct in the female?

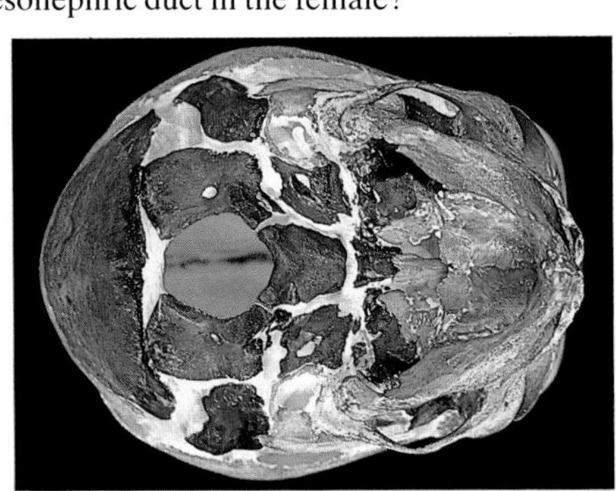

74 (a) Identify those bones which are formed in cartilage.
(b) What are the neurocranium and viscerocranium?
(c) Which parts of the viscerocranium are formed in membrane?

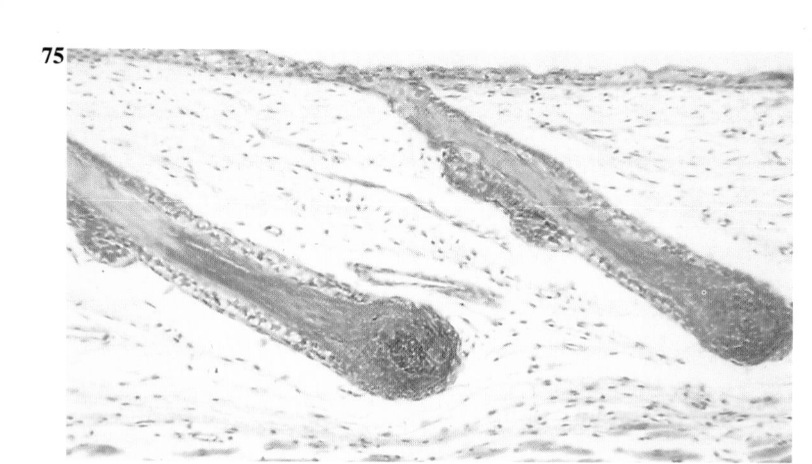

75 (a) Hair follicles begin as a solid downgrowth of which layer of the epidermis?
(b) What two types of hair are present in the fetus?
(c) How do most sebaceous glands develop?
(d) Name two regions where the sebaceous glands develop otherwise.

76 (a) How do early blood vessels form?
(b) Which vessel is formed when the two dorsal aortae caudal to the eleventh somite fuse?
(c) Which embryonic blood vessels form the common carotid arteries?

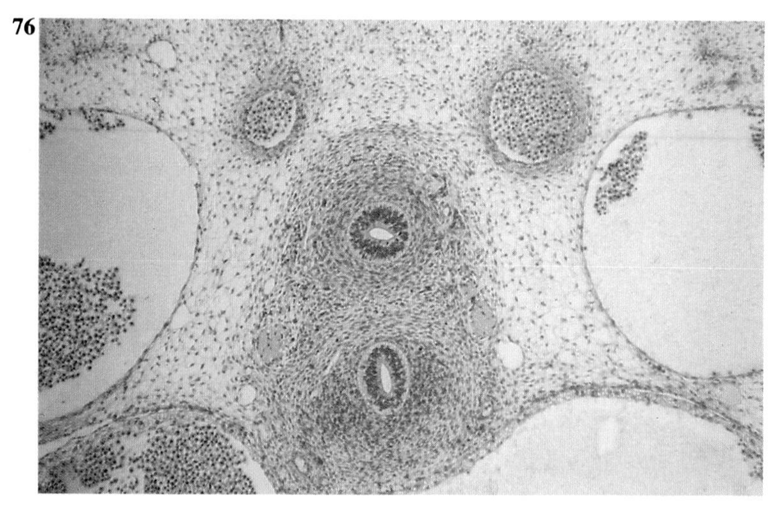

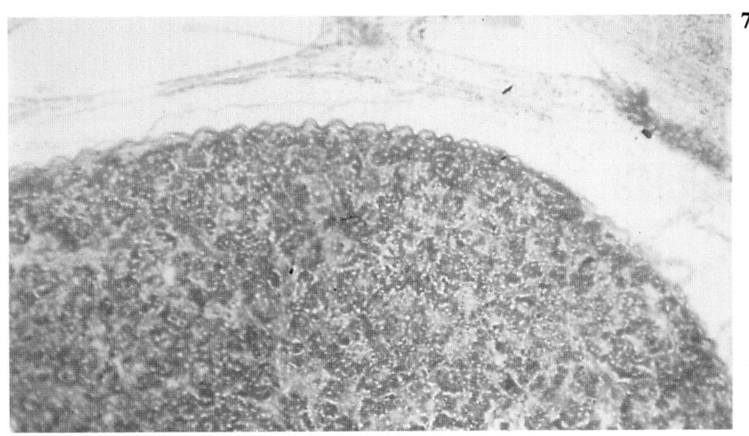

77 A coronal section of the hypophysis.
(a) Which hormones does it produce and when?
(b) Which appears first in the pars anterior: the acidophilic or basophilic granules?

78 (a) The dental laminae in each jaw form ten oval buds. Are these teeth the primordia for the deciduous or the permanent teeth?
(b) What are the three stages of tooth development?
(c) Describe the middle of the three stages?

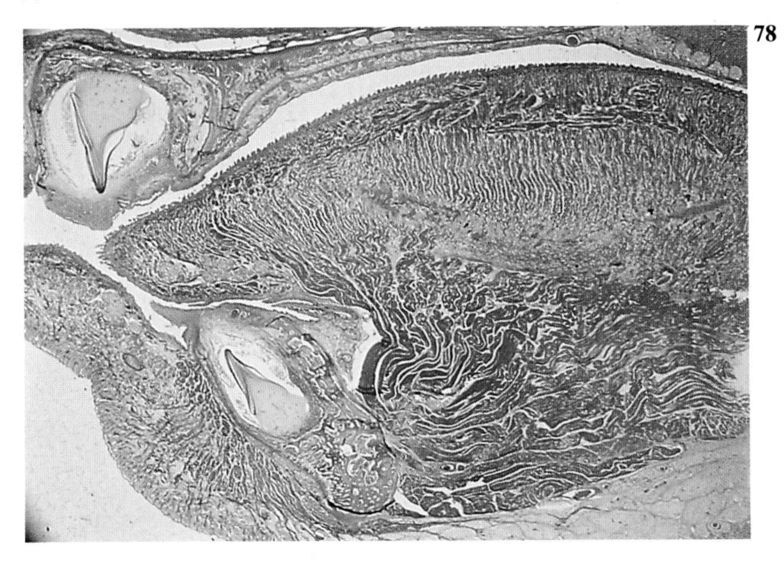

79

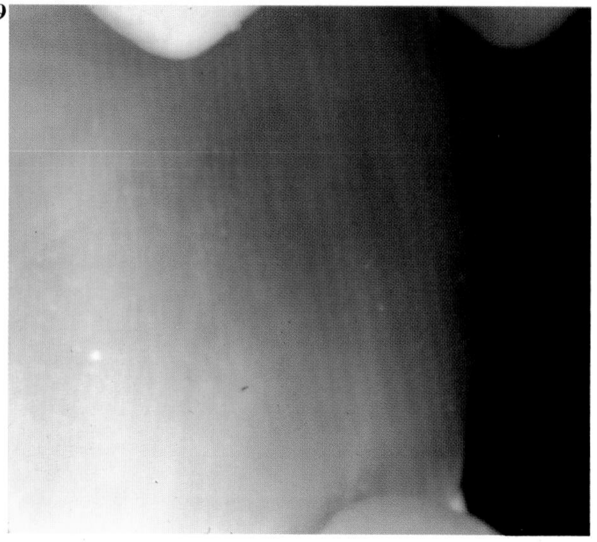

79 The thorax and abdomen of a 45mm CR fetus. Initially the body wall consists of a thin layer of epithelium and mesenchyme.
(a) What is the origin of the muscular layer in the anterior abdominal wall?
(b) What is the area of fusion between the two muscle sheets?

80

80 (a) How is the chorionic sac anchored to the maternal endometrium?
(b) What is the immediate fate of the decidua basalis?
(c) What is the ultimate fate of the decidua capsularis in the fetus?

81

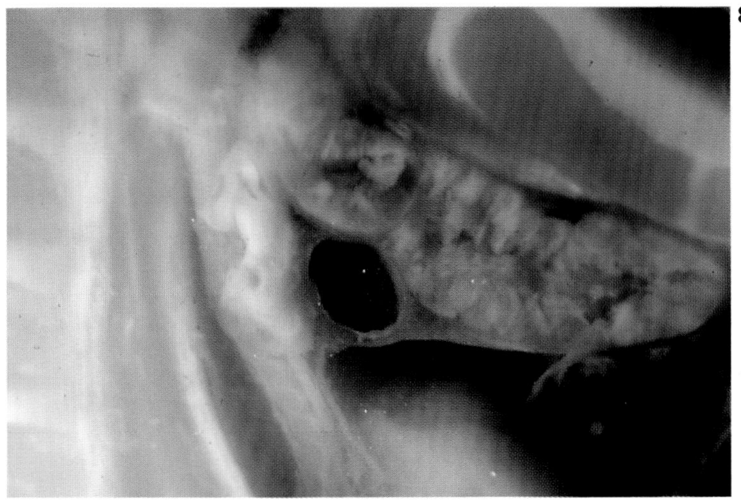

81 Sagittal section through the manubrium sterni (top right) and vertebrae (left) of a fetus.
(a) What is the normal shape of the thymus gland?
(b) Where does the thymus originate?
(c) What are the derivatives of the endoderm and mesenchyme in the thymus?
(d) When do lymphocytes from the thymus circulate in the blood stream?

82 The adult colliculi.
(a) Which tissue gives rise to the four colliculi in the tectum?
(b) Of which embryonic tissue is the substantia nigra a derivative?

82

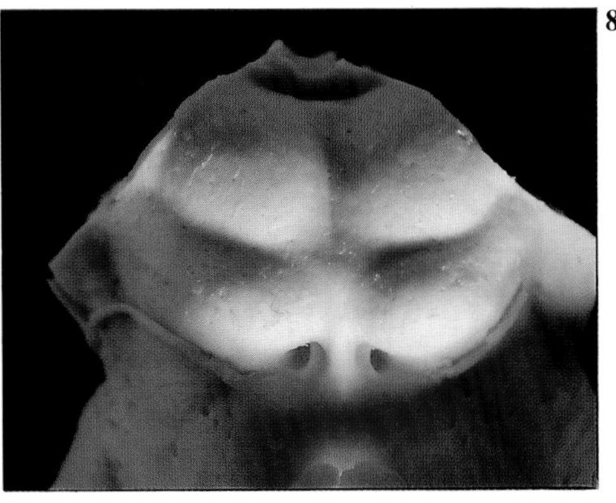

83

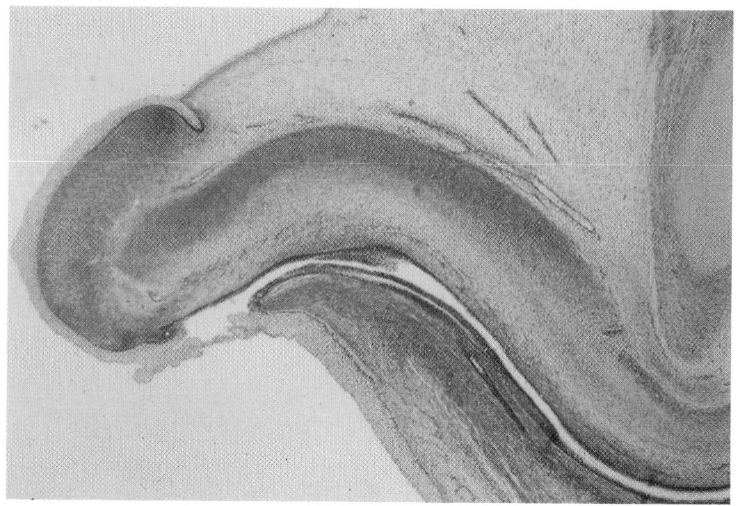

83 Longitudinal section of 70mm CR phallus.
(a) From which tissue does the epithelium of the penile urethra form?
(b) How does the glandular penile urethra form?
(c) From which tissue is the epithelium of the female urethra derived?

84 Transverse section of the head at the level of the hemispheres and hindbrain. The tubotympanic recess grows between the otic vesicle and the external acoustic meatus.
(a) The recess is a derivative of which pharyngeal pouch?
(b) What do the proximal and distal ends of the diverticulum (recess) form?
(c) The developing middle ear cavity surrounds which structures?

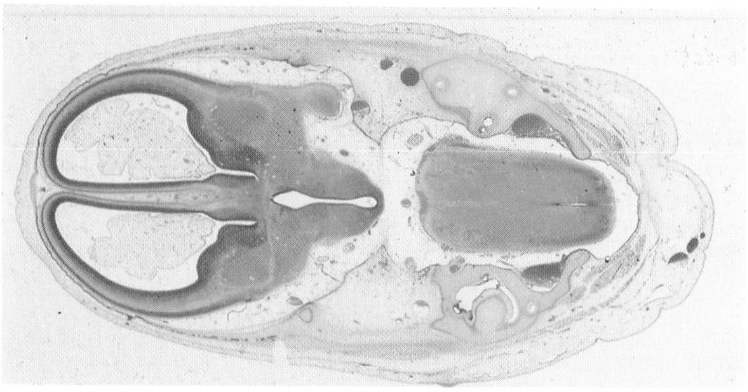

85 Compare the male and female fetuses.
(a) With which structure in the male are the labia majora homologous?
(b) With which structure in the male is the female urethra homologous?
(c) What happens to the primitive urethral groove in the female?

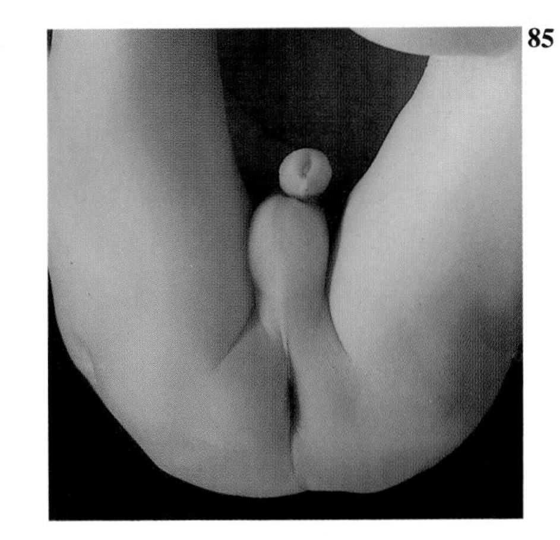

85

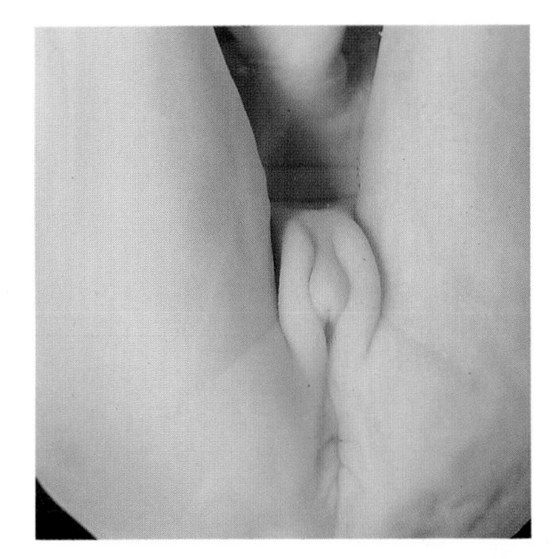

86 (a) What is the source of amniotic fluid in the early embryo and fetus?
(b) List the functions of the amniotic fluid.
(c) What are the constituents of fetal amniotic fluid?

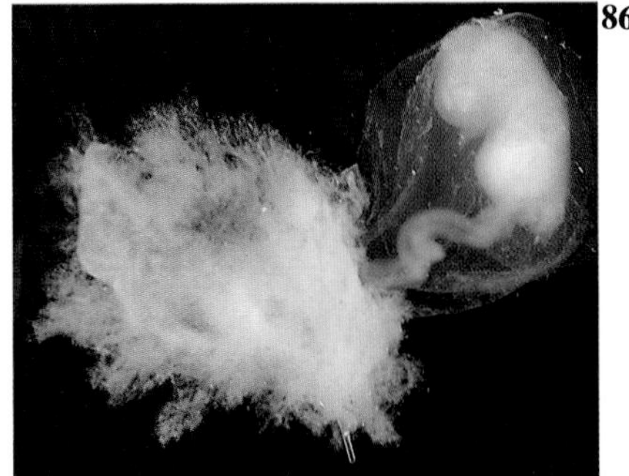

86

87

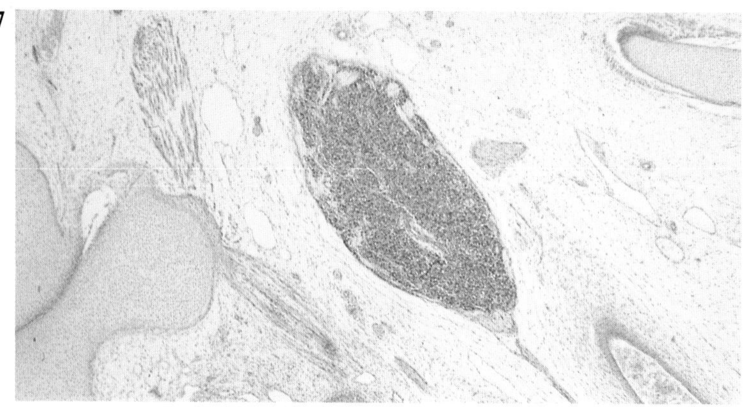

87 The tonsils form from many different primordia. This is the lingual tonsil from a 70mm CR fetus.
(a) How do the lingual tonsils arise?
(b) How do the palatine tonsils arise?
(c) How do the pharyngotympanic tonsils arise?
(d) How do the pharyngeal tonsils arise?

88 A scanning electron micrograph of fetal stomach.
(a) At what time do mucosal pits appear?
(b) The fetal stomach contracts during Week 11. Does true peristaltic activity occur?
(c) When is rennin present?

88

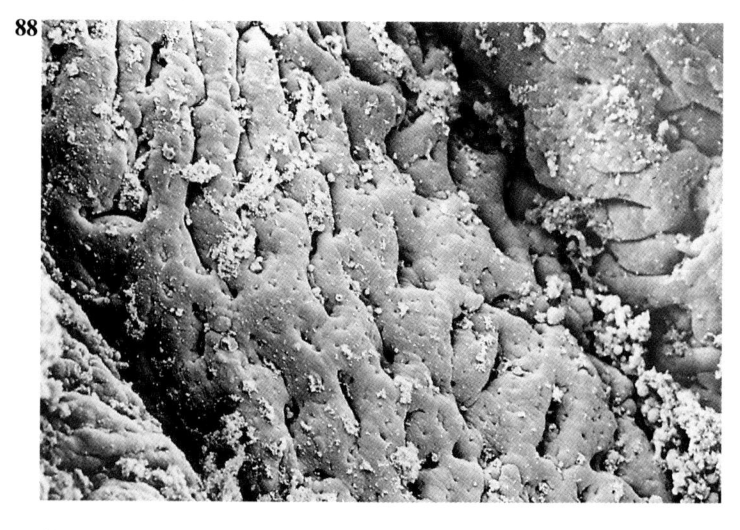

89

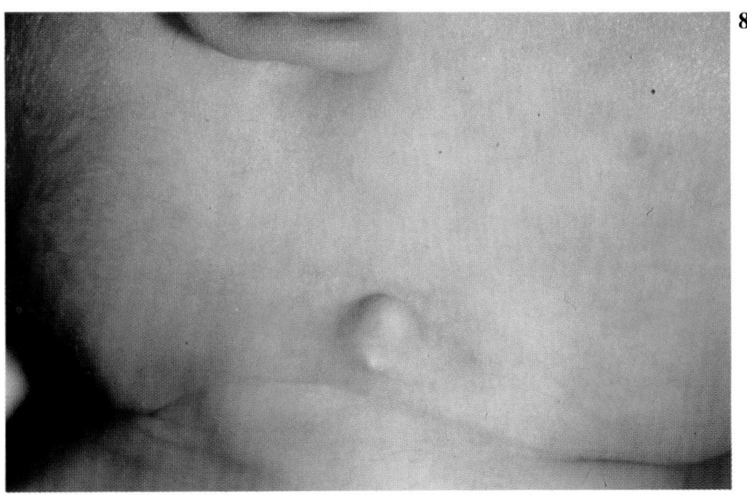

89 A branchial remnant is illustrated.
(a) Are branchial anomalies common or uncommon?
(b) What are the two main syndromes associated with the First Arch Syndrome?
(c) What is believed to be the cause of this?

90 A transverse section of lung in a 170mm CR fetus.
(a) What is the histological appearance at this stage of lung development?
(b) When are respiratory movements visible by ultrasound during the fetal period?
(c) What do these movements cause?

90

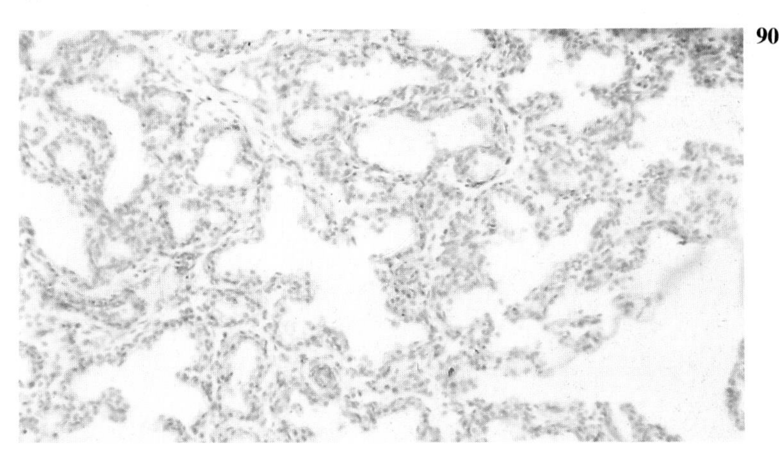

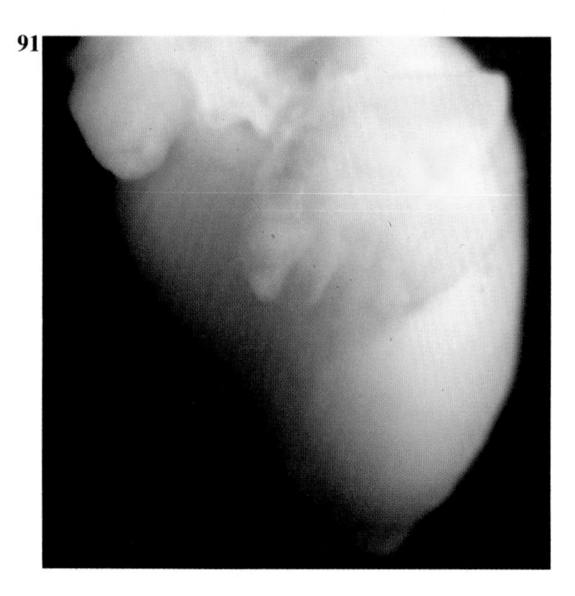

91 (a) When does the heart attain its definitive structure?
(b) The smooth, internal walls of the right atrium are derivatives of which embryonic structure?
(c) What forms the coronary sinus?

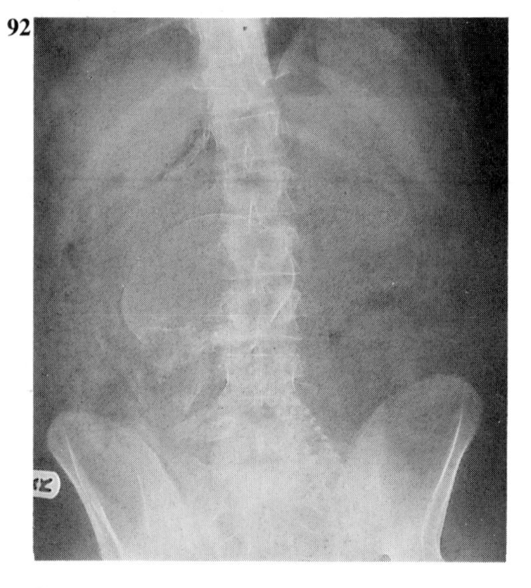

92 (a) Identify this type of presentation.
(b) Which imaging technique has routinely replaced this one?
(c) Why are X-rays of pregnant women no longer taken routinely?

93 Congenital pseudoarthrosis: a rare condition.
(a) What is the sensitive period of development for limbs?
(b) Is either oligohydramnios or polyhydramnios commonly associated with limb deformities?
(c) Which drug widely ingested during pregnancy as a sedative and antinauseant, caused other severe limb deformities?

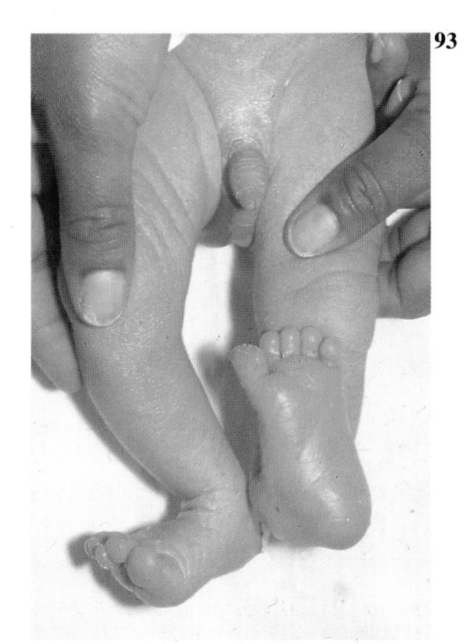
93

94 (a) What are the apparent differences between the colon shown and the adult colon?
(b) List the constituents of the colonic contents.
(c) Following the absorption of lipids, what term is used for the colonic contents?
(d) The presence of colonic contents in amniotic fluid is indicative of what condition?

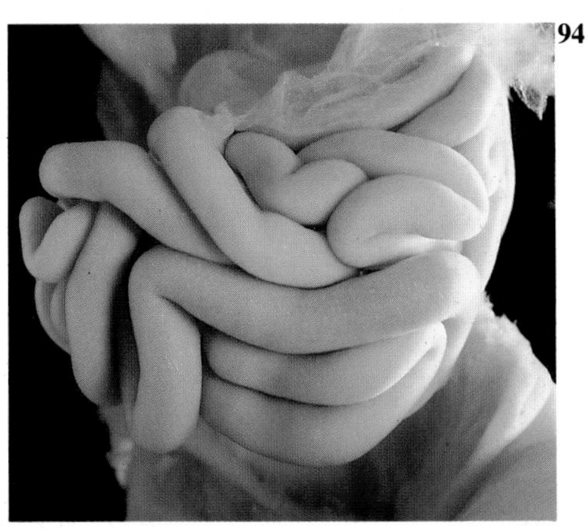
94

95

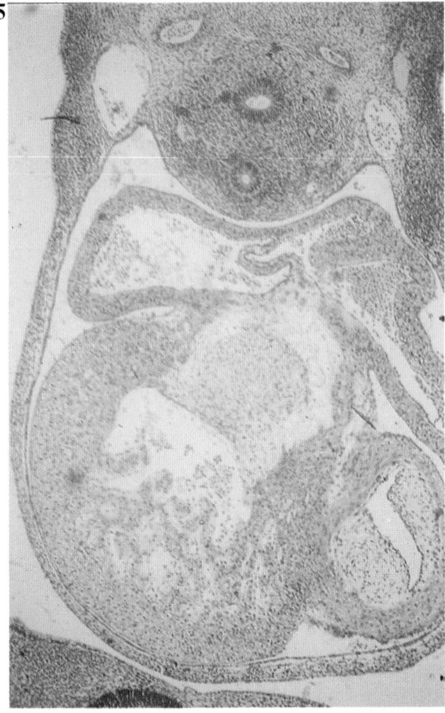

95 Horizontal section of the heart.
(a) How do the ventricular walls become trabeculated?
(b) From which tissue do atrioventricular valves form?
(c) The left sinoatrial valve fuses with the septum secundum. Into which structure are they incorporated?
(d) Which ventricle is thicker at birth: right or left?

96

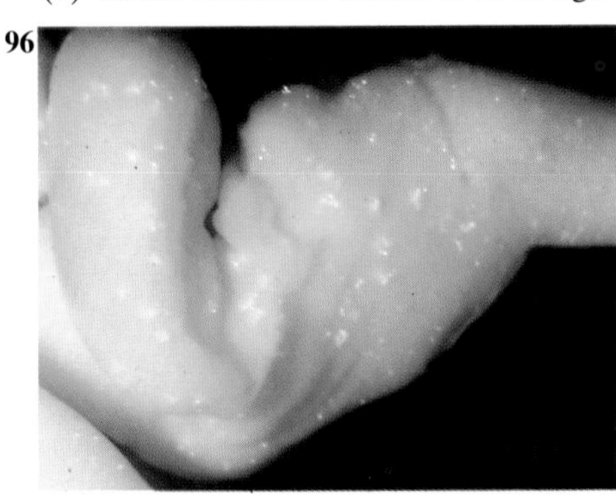

96 (a) From what structure is the developing cerebellum derived?
(b) As it grows, which two structures does it overlap?

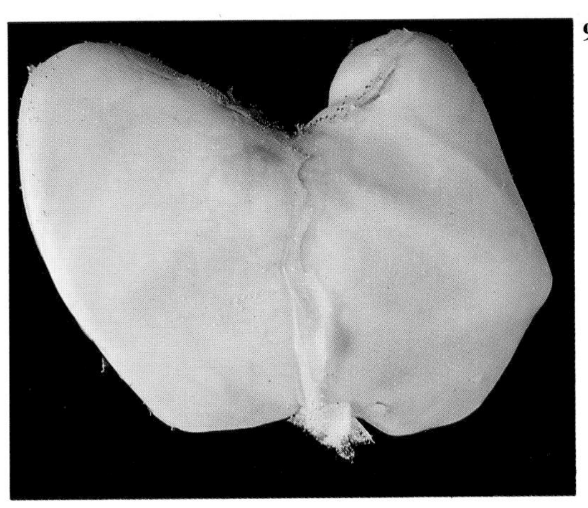

97 The liver of a 101mm CR fetus viewed from the ventral surface.
(a) How does this organ arise?
(b) Name the derivatives of the mesenchyme.
(c) When and in conjunction with what other organ does haemopoiesis occur in the liver?

98

98 (a) Identify the gland taken from a 57mm CR fetus.
(b) Which adjacent tissue contributes neuroectoderm to the medulla?
(c) When does the zona reticularis form?

99

100

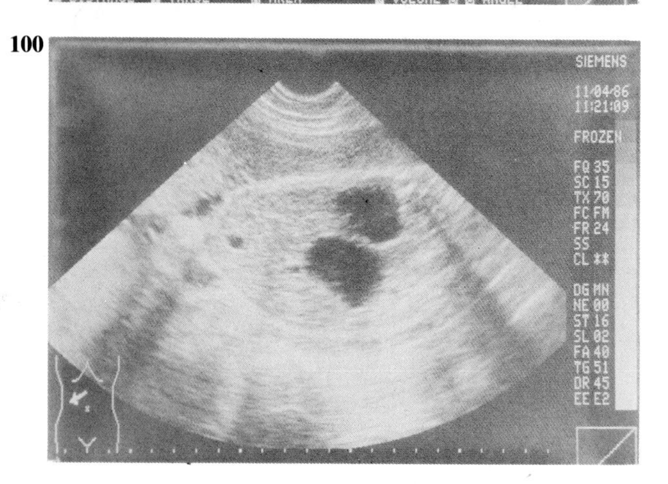

99, 100 Ultrasound scans showing bilateral hydronephrosis: both transverse and longitudinal views.

(a) Which other organ is visible?

(b) Is the fetus usually male or female?

(c) Would any abnormality of the amniotic fluid be present?

101 Imperforate anus in a newborn.
(a) How common are anomalies of the anus and rectum?
(b) In higher levels of rectal agenesis in the male a fistula may be present in 70 per cent of cases. What other structure does this connect with?
(c) What associated anomalies may be seen in X-rays?

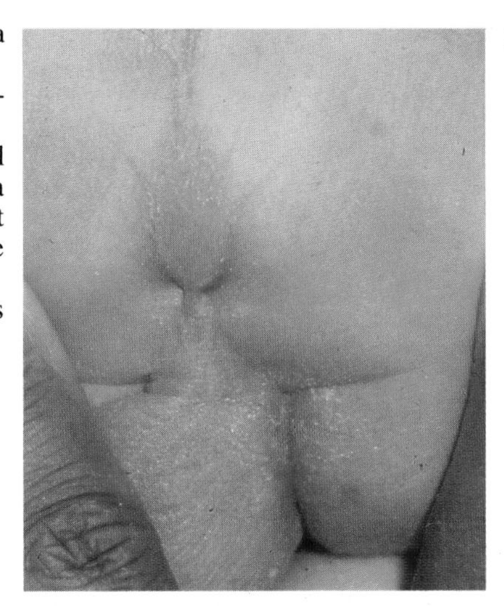

10

102 A transverse section through a Stage 7 embryo.
(a) At what stage of development is this bilaminar embryo?
(b) Identify the two primary embryonic layers.
(c) Which primary layer gives rise to the nervous system?

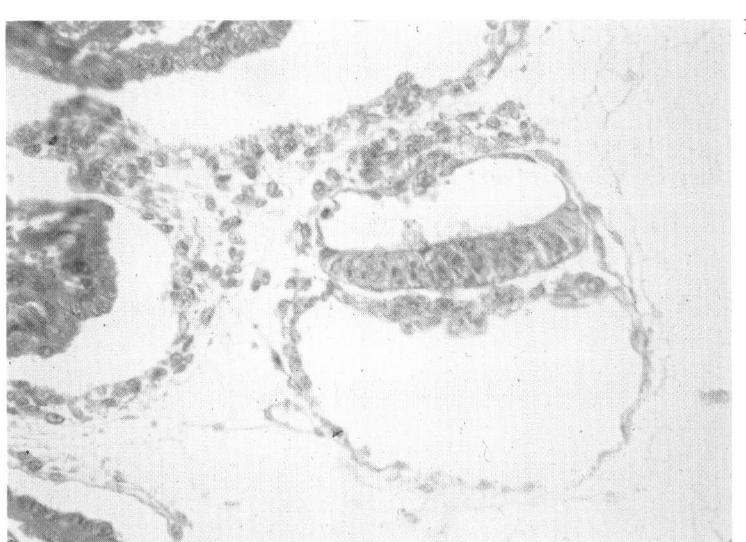

102

103

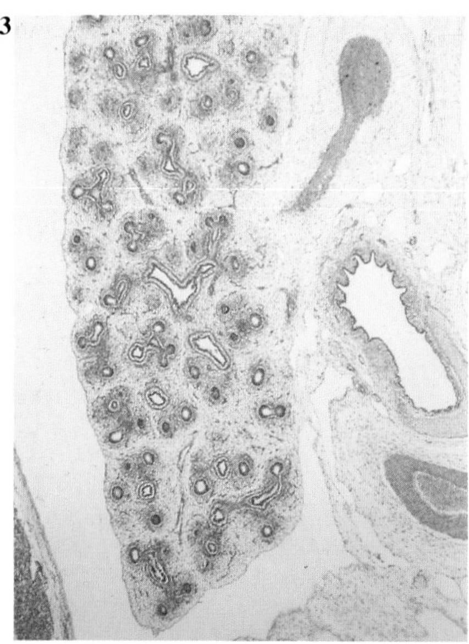

103 Transverse section of a 70mm CR fetal lung.
(a) Identify this histological stage of lung development.
(b) The pleura form from which cells?
(c) From which embryonic layer do the bronchial cartilagenous plates or rings develop?

104 Hemisection of adult testis.
(a) What do the seminiferous cords form in the male?
(b) What is the mesentery that suspends the testis?
(c) Of what two types of cells are the walls of the seminiferous tubules composed?

104

105

105 The right leg bud of a 12mm CR (Stage 16) embryo.
(a) Identify the role of the thickened ectodermal area at the tip of the leg bud.
(b) List the derivatives of the ectoderm.
(c) List the derivatives of the mesenchyme.

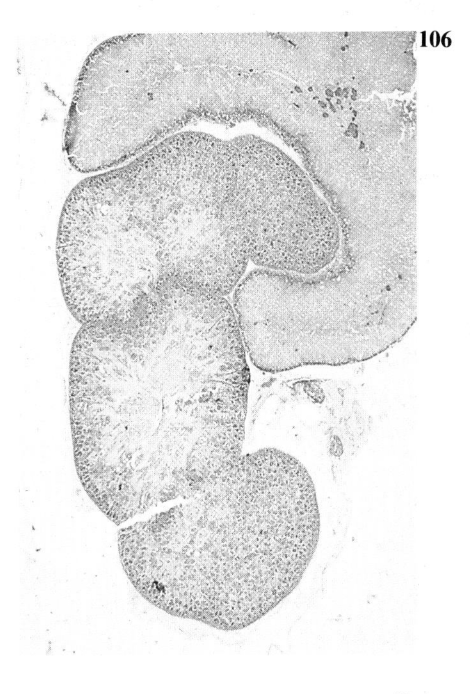

106

106 (a) Identify the gland in the upper part of this section.
(b) What is its dual origin?
(c) How does the adult arrangement of cortex and medulla form?
(d) Which fetal anomaly is the usual result of hyperplasia of the fetal cortex?

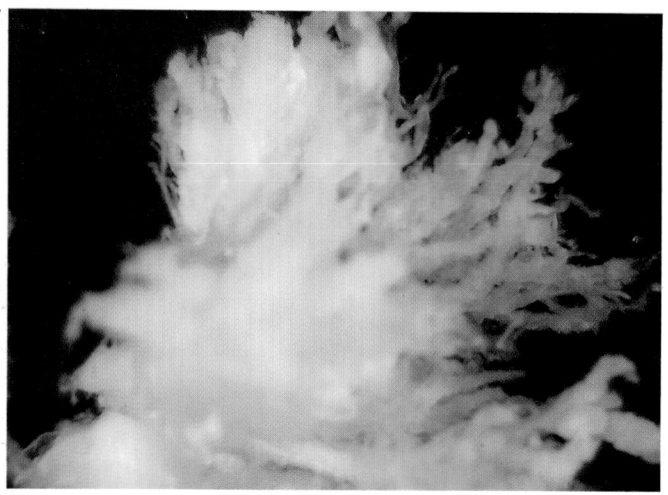

107 (a) The primary chorionic villi are composed of what two main cell types?
(b) What are secondary villi?
(c) What are tertiary villi?

108 The thyroid of a 51mm CR fetus.
(a) Name its site of origin on the tongue.
(b) What is the fate of the stalk of the thyroid?
(c) Remnants of the thyroglossal duct can give rise to what?

108

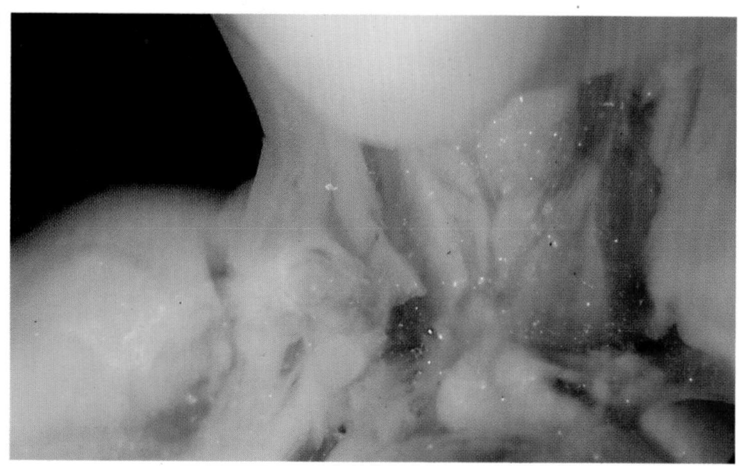

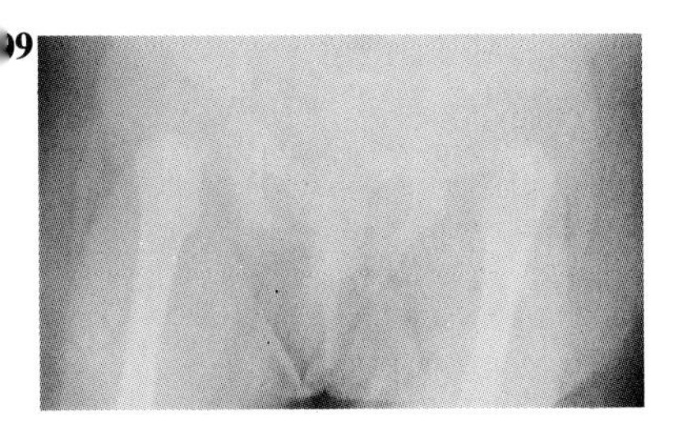

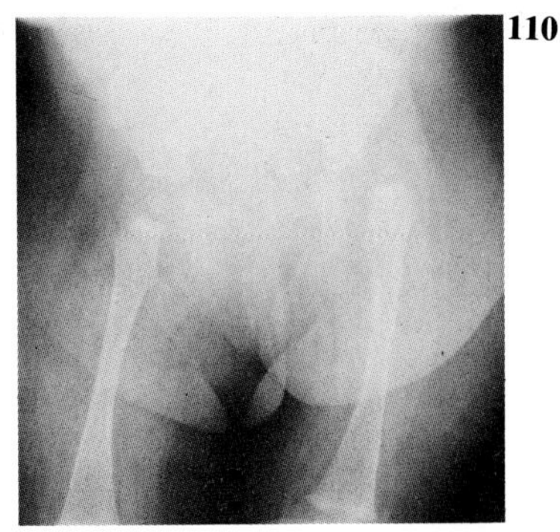

109, 110, 111 (a) What condition is illustrated?
(b) What is its frequency?
(c) Name two causative factors.

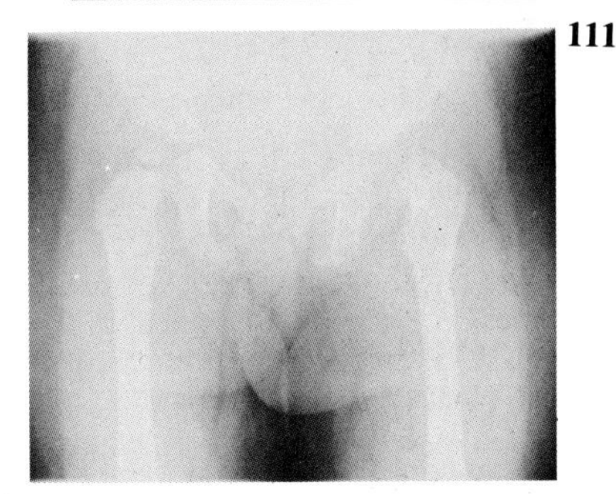

112 Thyroid of a full-term fetus.
(a) When does colloid normally appear?
(b) The ultimobranchial body which fuses with the thyroid is a derivative of which primary embryonic tissue?
(c) Which tissue is derived from the ultimobranchial body?

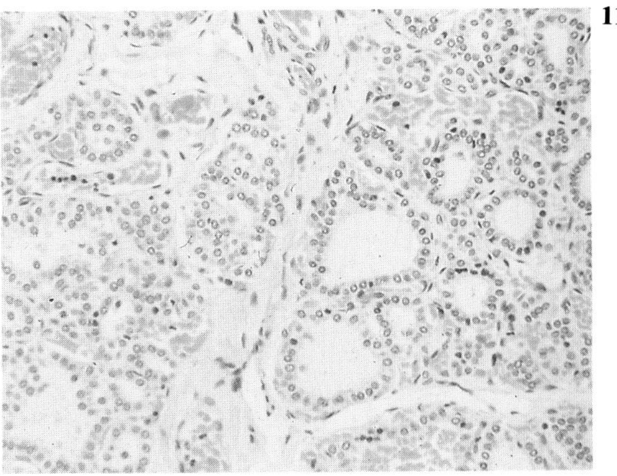

113

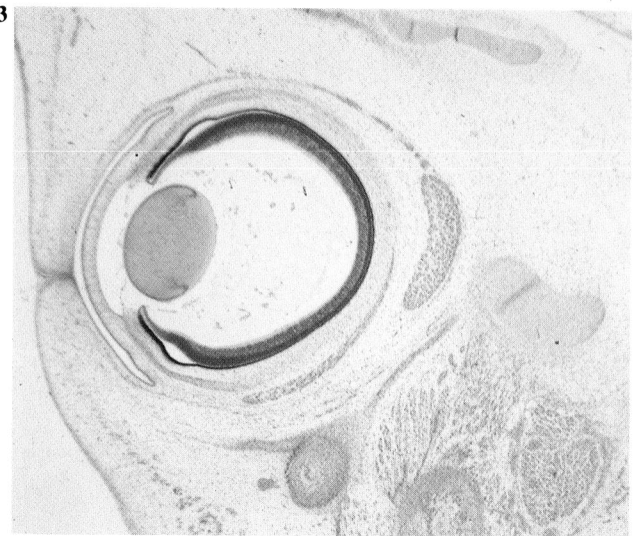

113 A longitudinal section of the eye in a 40mm CR fetus.
(a) The choroid and sclera are derivatives of which tissue?
(b) How does the pupillary membrane form?
(c) What is the origin of the lacrimal glands?
(d) What is the origin of the nasolacrimal duct?

114 (a) What is another term for an undescended testis?
(b) When do the testes normally descend?
(c) How do they descend?
(d) What is the processus vaginalis?

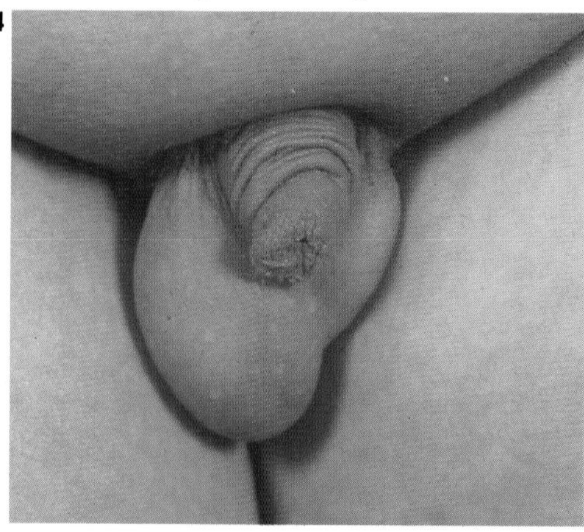

115

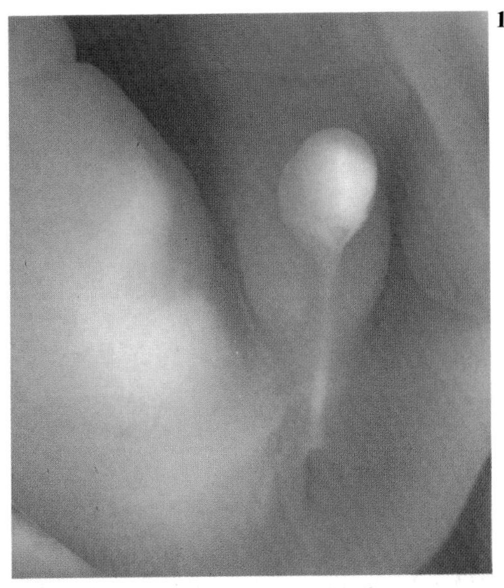

115 The male external genitalia.
(a) Which structure is formed in the male by the fusion of the genital (labioscrotal) swellings?
(b) From which tissue do the corpora cavernosa of the penis and corpus cavernosum urethrae form?
(c) How does the prepuce form?

116

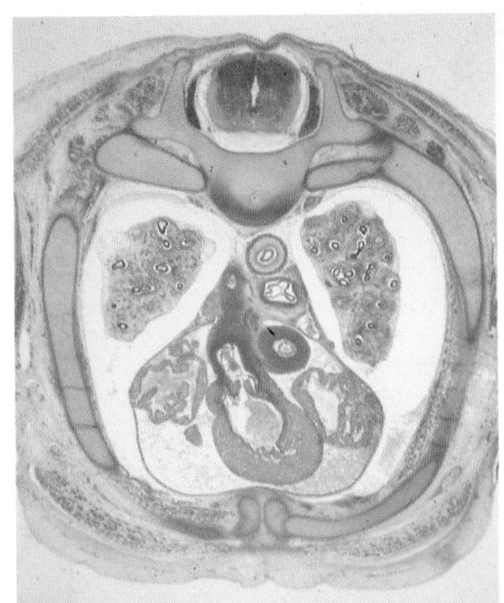

116 Transverse section through the thorax of a fetus.
(a) When are the pericardial and pleural cavities separated?
(b) The pleuropericardial membranes fuse with which tissue to separate the two cavities?
(c) Does the right or left pleuropericardial opening close first?

117

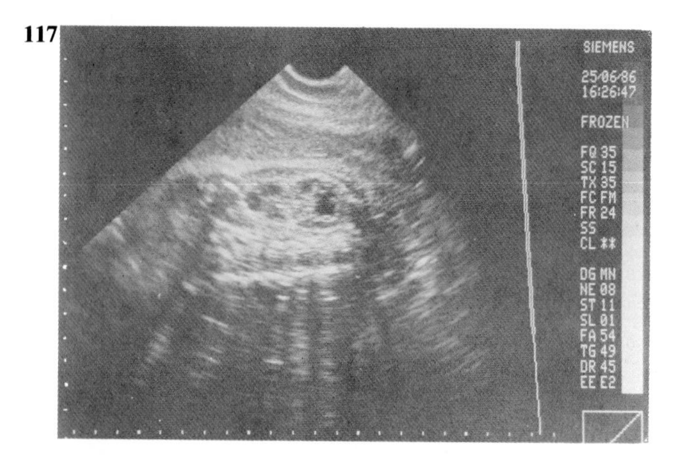

118

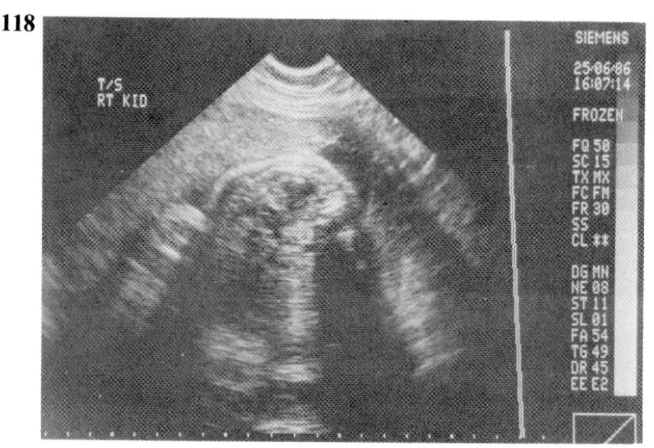

117, 118 Pelvi-ureteric junction obstruction (PUJO): longitudinal and transverse views in ultrasound scans.

(a) What is the condition of the renal pelvis?
(b) Is an adequate amount of amniotic fluid present in this condition?
(c) How much amniotic fluid is normally present by Week 37?
(d) What is the turn-over rate of water in amniotic fluid in the late fetus?

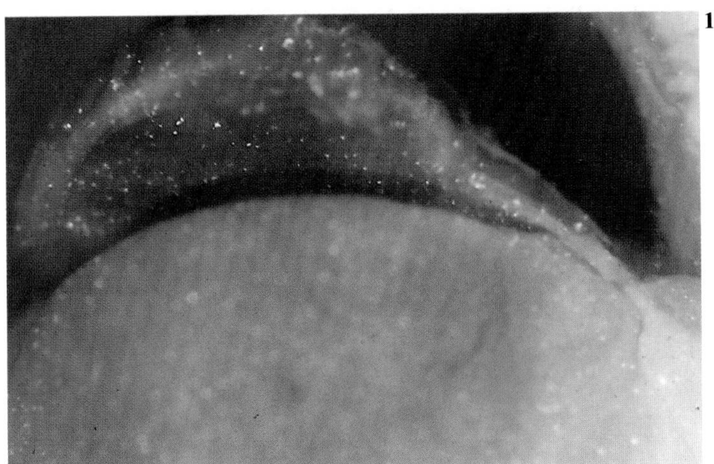

119 (a) The septum transversum shown in this early embryo will form which part of the diaphragm?
(b) Give three other contributions to the diaphragm.
(c) Which vertebral levels contribute myotomes via the pleuropericardial membrane to the diaphragm?

120 Right side of the heart of a fetus.
(a) Which part of heart acts as a temporary pacemaker?
(b) Why does the interventricular groove form?

120

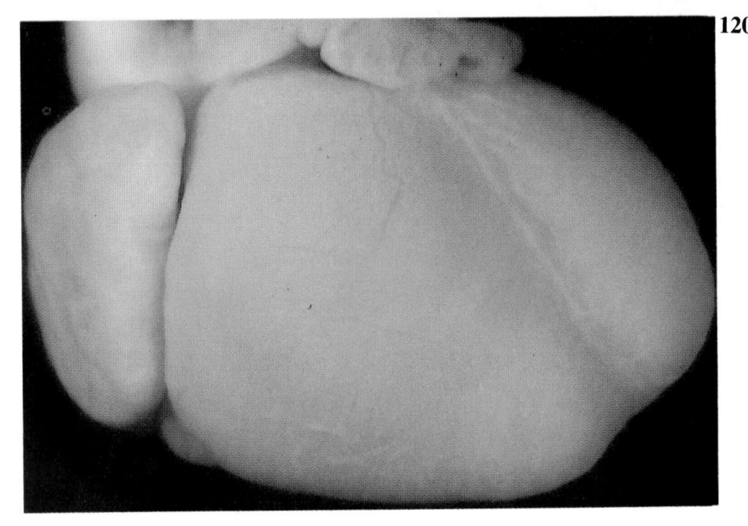

121

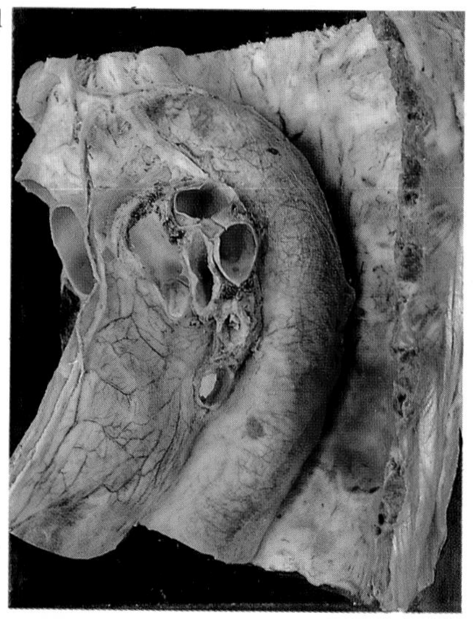

121 The adult aorta viewed *in situ* from the left side.
(a) Identify the embryonic origin of the three components of the adult aorta.
(b) The two pulmonary arteries supplying the lungs were branches of which branchial arch arteries?
(c) Which embryonic blood vessel forms the ductus arteriosus?

122 A section of the heart in a 20mm CR embryo.
(a) At birth, which gland covers the anterior surface of the right atrium?
(b) Which early valves direct blood from the sinus venosus as it enters the atrium?
(c) How does the septum spurium form?

122

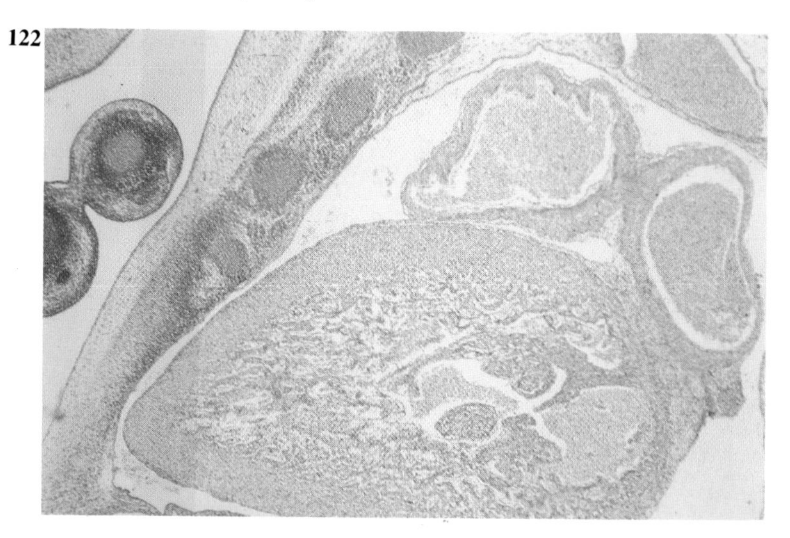

123 (a) Define a dermatome.
(b) Axial lines may be used clinically because they illustrate which phenomenon?
(c) Which are the palmar dermatomes?
(d) Which are the plantar dermatomes?

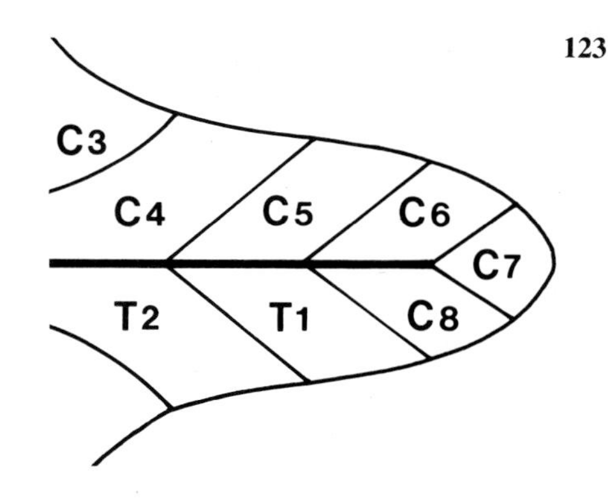

124 The brain of a fetus viewed from the left side.
(a) When do the sulci and gyri develop?
(b) Which sulci develop first?
(c) What function are gyri and sulci believed to have?

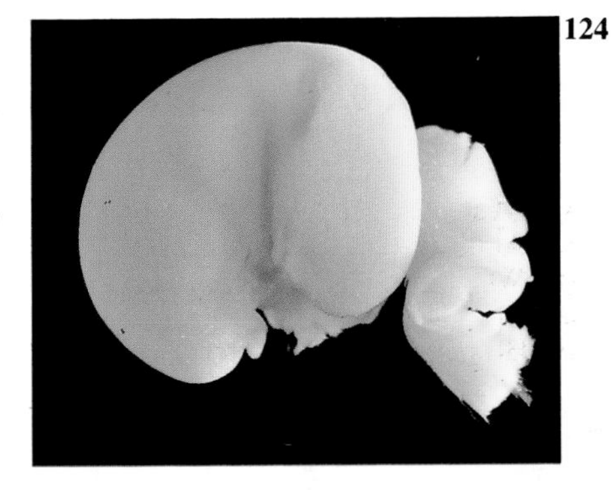

125 The face of an early embryo viewed from the front and right side.
(a) What are the five fused processes bounding the early mouth?
(b) What is the depression on the surface ectoderm marking the site of the future mouth?
(c) Give three reasons why the neonate's face is small.

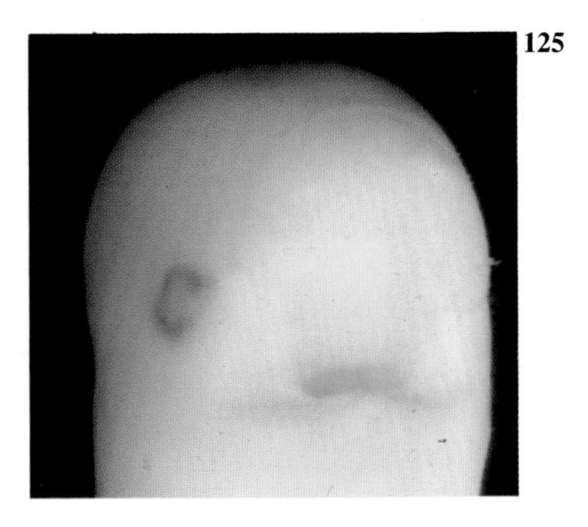

126

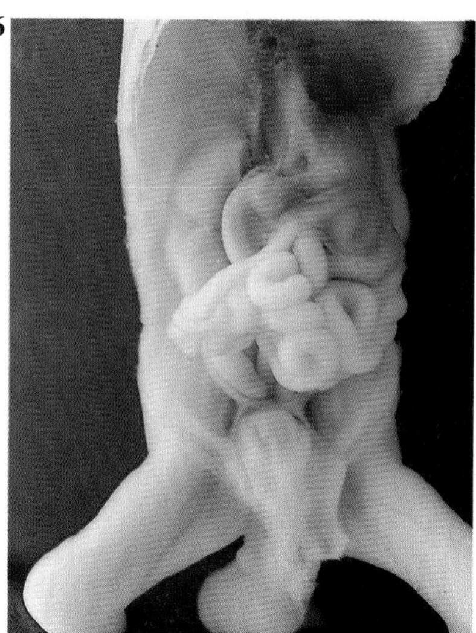

126 Following midgut rotation and reduction the hindgut moves to the left. Note the vermis appendix.

(a) What are the derivatives of the hindgut?

(b) What is the arterial supply of the hindgut?

(c) What happens to the mesenteries of the hindgut?

127 Horizontal section of the tongue and maxilla.

(a) What is the origin of most of the tongue muscles?

(b) When do the papillae develop?

(c) What is the nerve supply to the anterior two-thirds of the tongue?

127

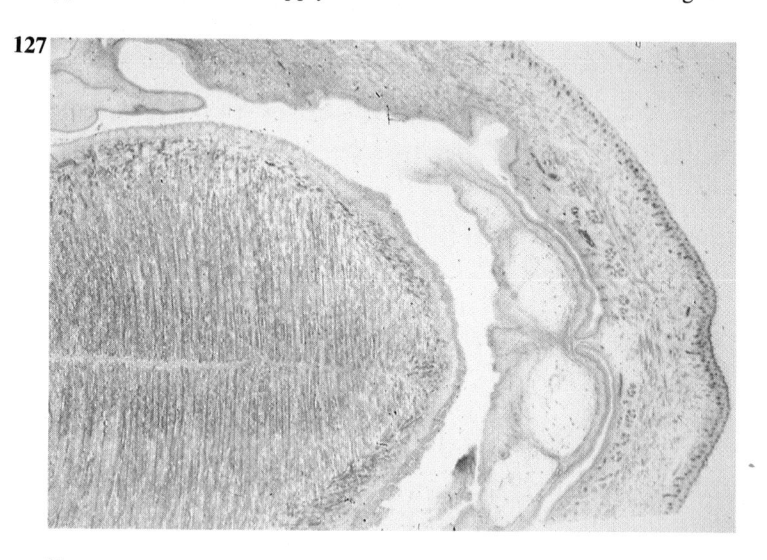

128 The uterus dissected open to display a 20mm CR (Stage 19) embryo.
(a) Name the fetal membranes formed by the zygote.
(b) Which fetal and maternal tissues form the placenta?
(c) What are the functions of the membranes and placenta?

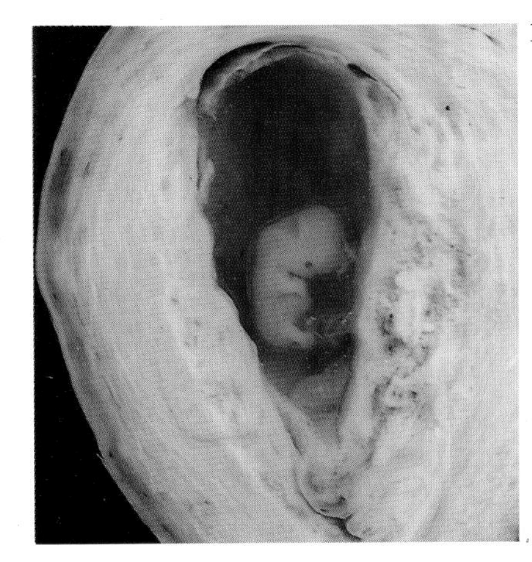

128

129 Horizontal section through the back and leg bud of a 10mm CR embryo.
(a) From which tissue do most skeletal trunk muscles primarily form?
(b) All muscles except one form from myoblasts which have differentiated from mesenchyme. What is the exception?

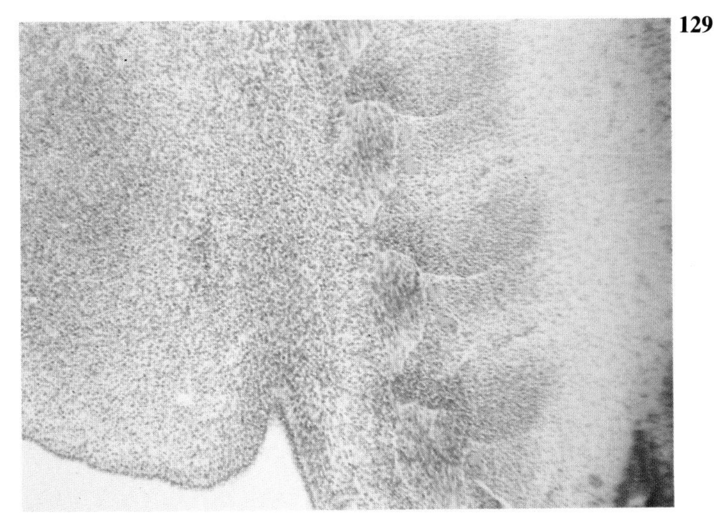

129

130

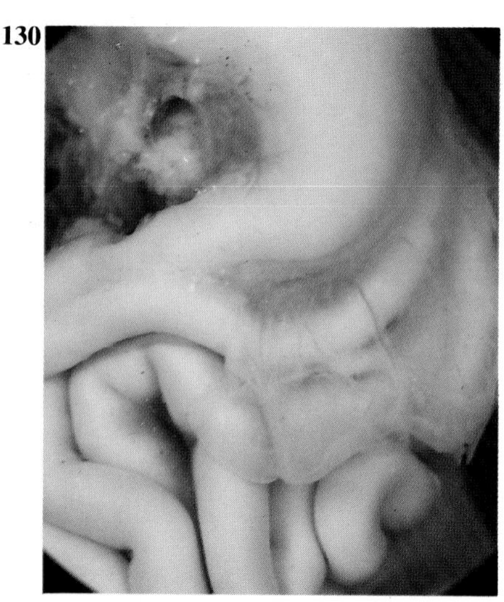

130 The stomach of a 92mm CR fetus and some of its associated mesenteries viewed from the front.
(a) Which adult structure will the dorsal mesogastrium form?
(b) Name the entrance to the lesser sac.

131

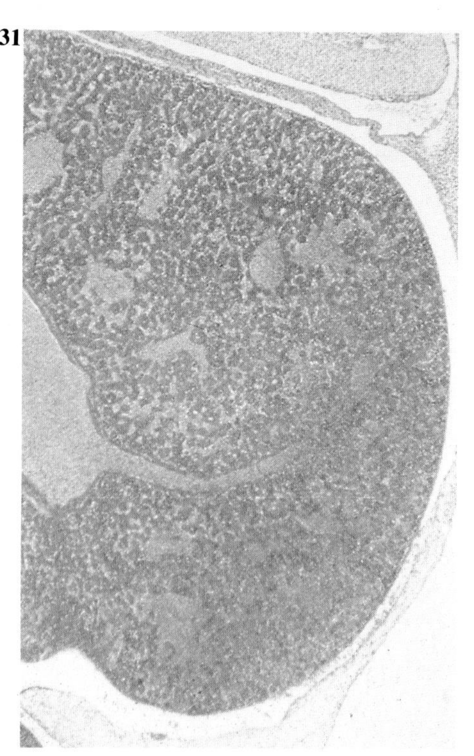

131 Parasagittal section through the liver of a 20mm CR embryo.
(a) How do the sinusoids form in this organ?
(b) When is bile produced?
(c) When are clotting factors produced in the liver?

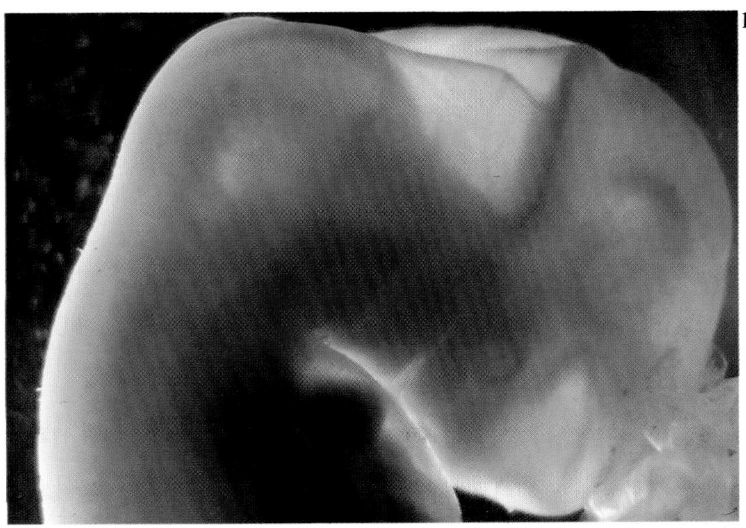

132 The brain of a 12mm CR (Stage 16) embryo viewed with transmitted light.
(a) In what order do the flexures appear?
(b) Which flexures eventually disappear?
(c) The walls of the rhombencephalon become splayed by which flexure?

133 An ultrasound scan of the fetal spinal column.
(a) What condition is illustrated?
(b) Give a general description of this condition.

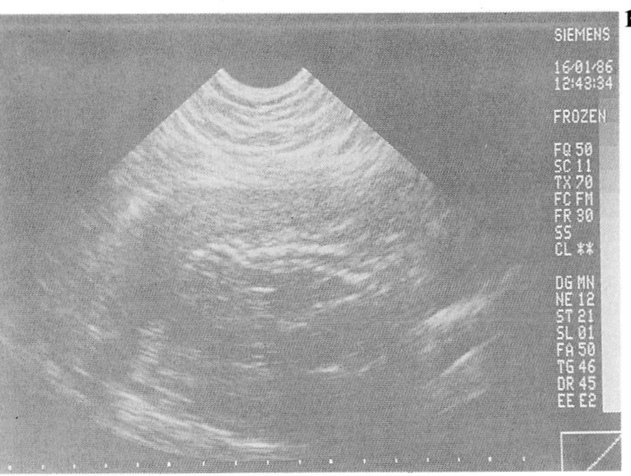

134

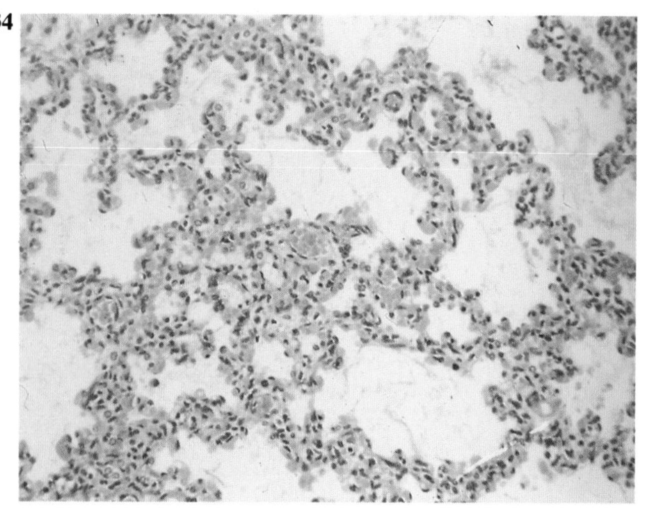

134 The terminal sac period of lung development.
(a) When does cell type I appear?
(b) Which cell type produces surfactant?
(c) What is the role of surfactant?
(d) Are almost all the alveoli present at birth?

135

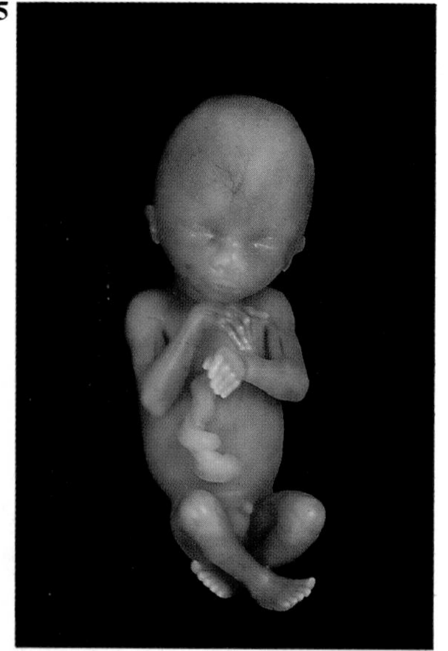

135 (a) How is the cord stump used medico-legally?
(b) What happens to the intra-abdominal portions of the umbilical arteries and vein at birth?
(c) Which blood vessel forms the adult ligamentum teres?

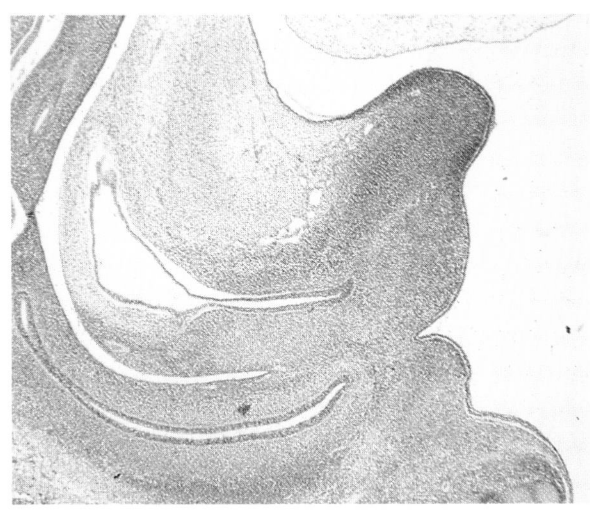

136

136 The genital tubercle elongates to form a phallus.
(a) What is the projection it carries from the urogenital sinus?
(b) When the urogenital membrane breaks down, which two areas become continuous?
(c) When the urogenital folds fuse in the male, what is the result?

137 These adult bones displayed are first and second arch derivatives.
(a) What are the derivatives of Meckel's (first arch) cartilage in the middle ear?
(b) What are the derivatives of Reichert's (second arch) cartilage in the middle ear?
(c) Which muscles form from the first and second arch mesoderm in the middle ear?
(d) In the late fetus the tympanic cavity expands and gives rise to what structure?

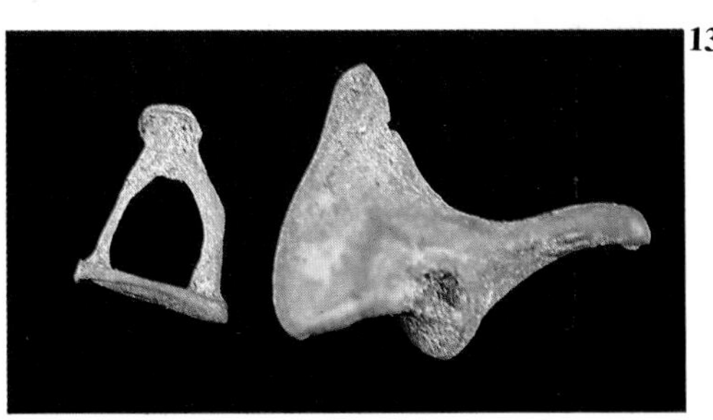

137

138

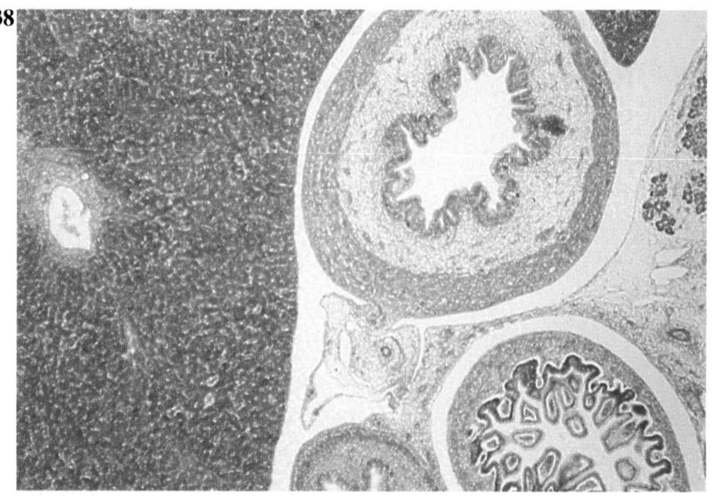

138 Transverse section of the stomach and liver of a 70mm CR fetus.
(a) How does the stomach form?
(b) During stomach rotation the left side of the stomach becomes which surface?
(c) The ventral mesogastrium exists on which fetal structures?

139 A transverse section of the early heart.
(a) What is the structure on the left?
(b) What septum will these ridges form?

139

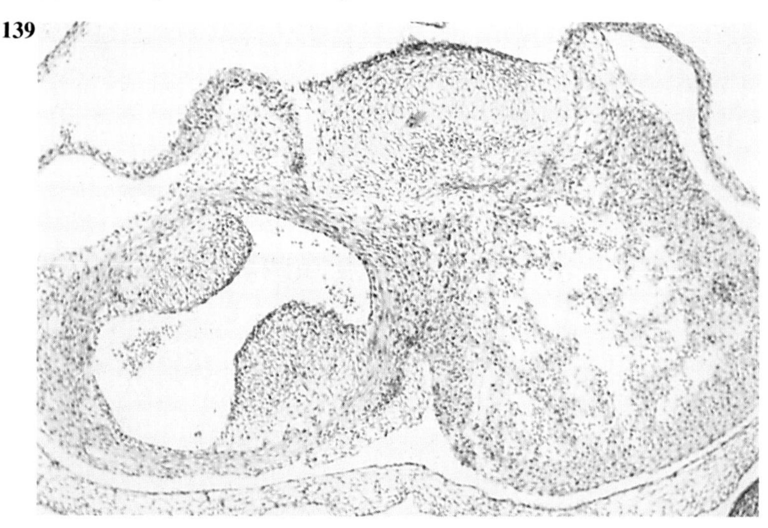

140 The early neural plate (Stage 10) embryo.
(a) What is the rudiment of the central nervous system?
(b) When does it appear?
(c) How does the neural tube form?

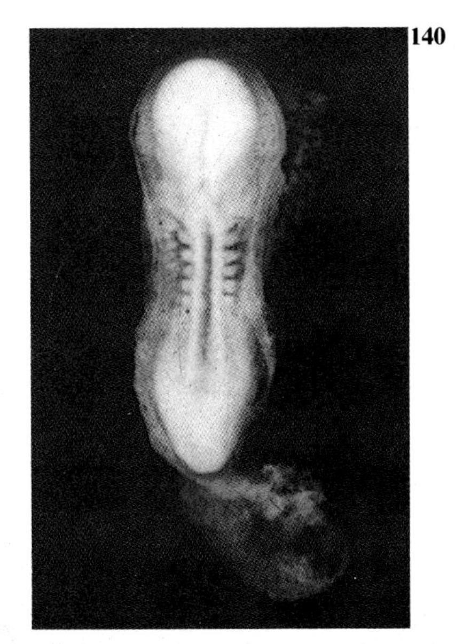
140

141 The cochlea and the eighth nerve. The middle ear is seen towards the left side.
(a) The upper and lower parts of the otic vesicles (otocysts) enlarge to form two areas. Which one will form the cochlea?
(b) What forms the constriction between the two parts?
(c) How do the semicircular ducts form?

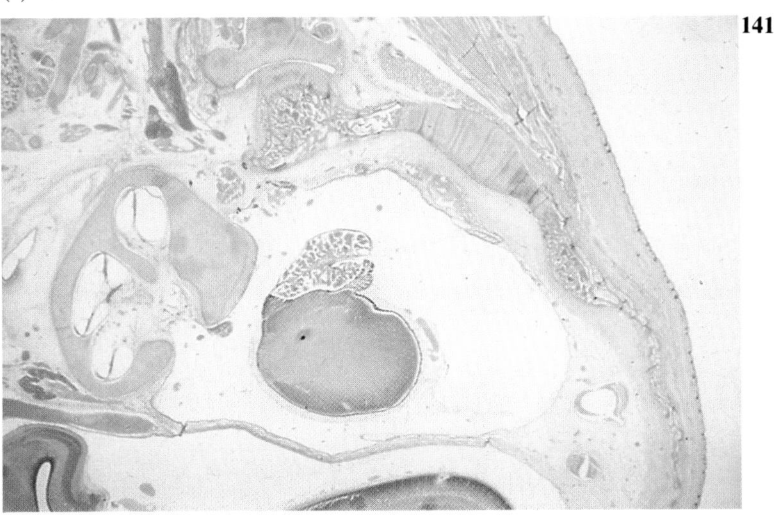
141

142

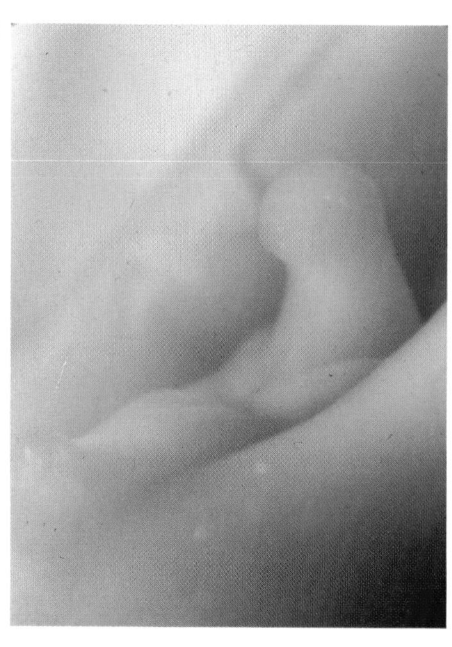

142 The external genitalia of a 45mm CR fetus viewed from the left.
(a) Until what age are the external genitalia of the two sexes similar in appearance?
(b) What adult structures form from the genital tubercle?
(c) The cloacal membrane is surrounded by two paired structures on each side. What are they?

143 (a) When does implantation occur?
(b) What is a decidual reaction?
(c) Describe the three regions of decidua in relation to the implantation site.

143

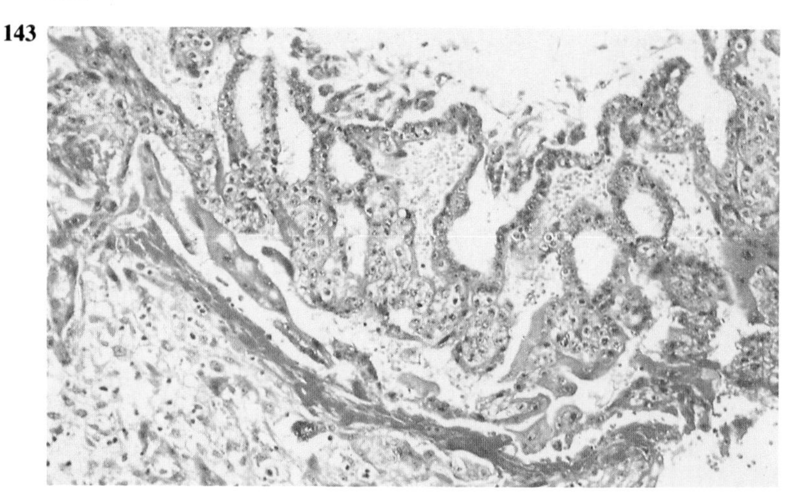

144 (a) Describe where the bone originates in the fetal palate.
(b) The high position of the neonate's epiglottis allows it to contact which structure?

144

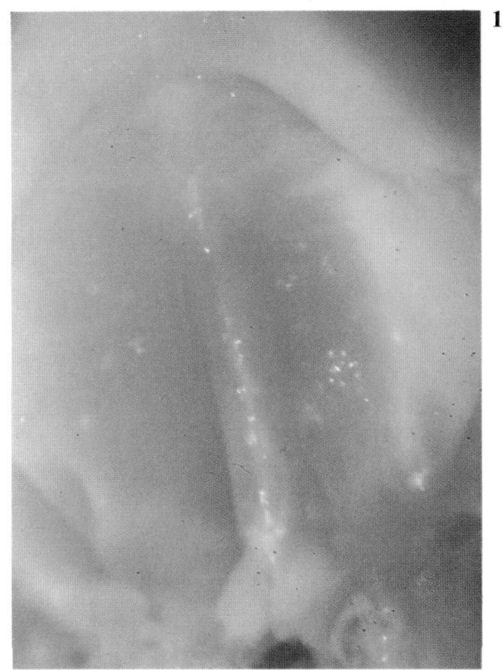

145 (a) A ring can be seen on the temporal bone. What membrane is normally present here?
(b) When does the mastoid process develop?
(c) What anatomical relationship is of importance in a forceps delivery?

145

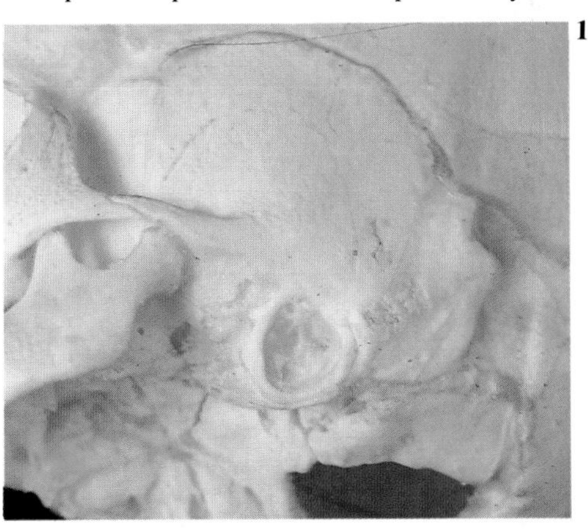

146

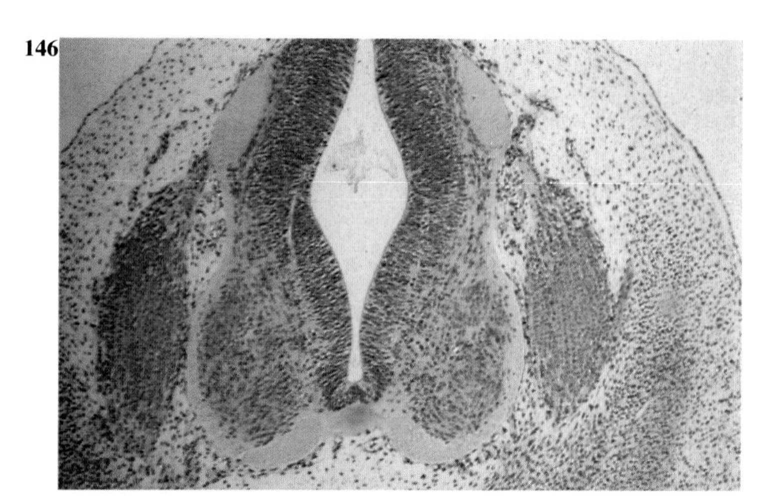

146 A transverse section of spinal cord in a 10mm CR embryo.
(a) How does the dorsal sensory nerve root form?
(b) Which adult structures will the basal lamina form?
(c) At which vertebral level does the neonate's spinal cord usually terminate?

147 The arm bud of an embryo viewed from the right.
(a) What are the cranial and caudal borders also called?
(b) Why is the cephalo-caudal maturity gradient of the upper limb retained throughout fetal development?
(c) How does the arm bud rotate to its adult position?

147

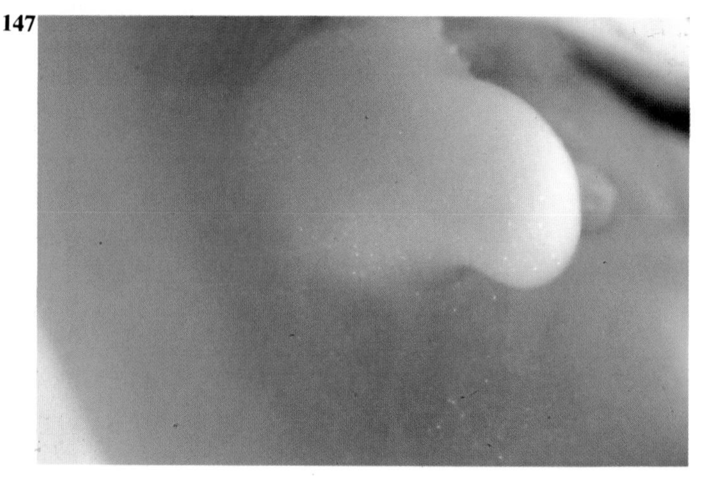

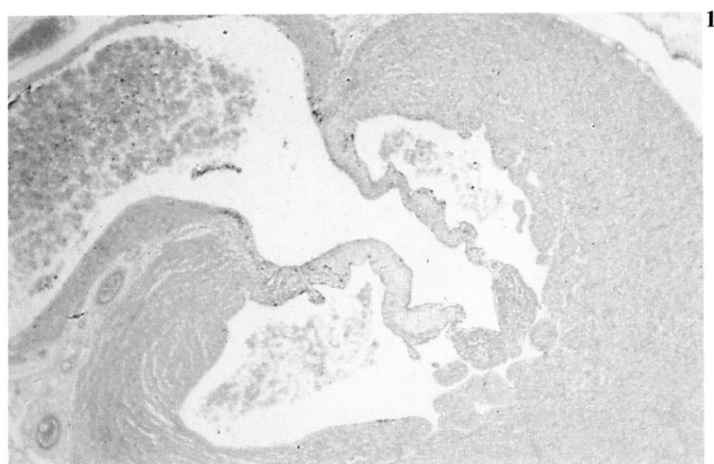

148 Ventricle of a 70mm CR fetus.
(a) The aorticopulmonary septum divides the bulbus cordis and truncus arteriosus into two vessels. What are they?
(b) What does the bulbus cordis form in the adult right ventricle and in the adult left ventricle?

149 The legs and feet of an embryo viewed from the right side.
(a) Are the longitudinal and transverse arches of the foot present in the newborn?
(b) When do epidermal ridges form on the plantar surface of the foot?

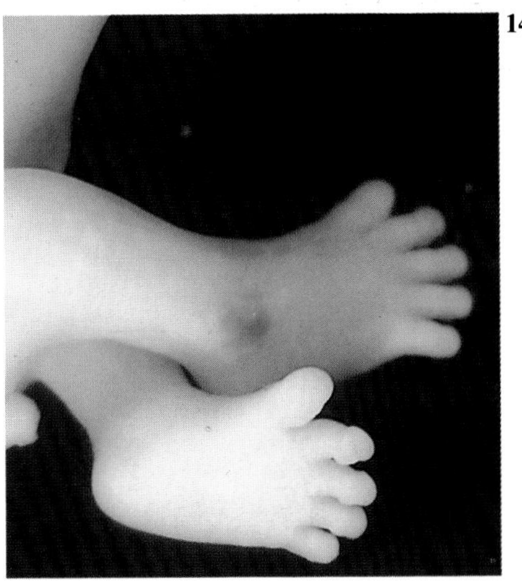

150

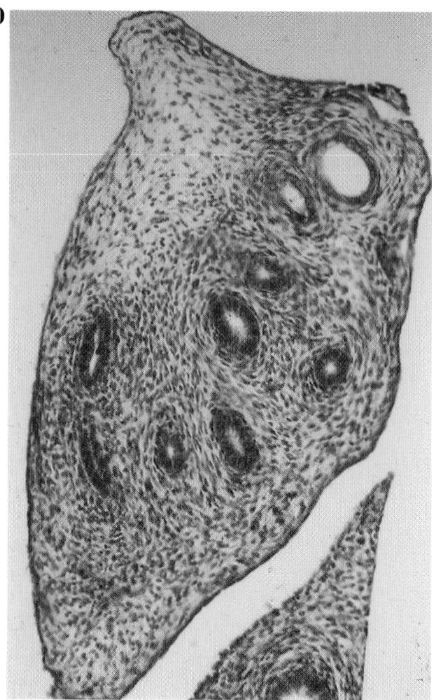

150 A longitudinal section of prostate in a 70mm CR fetus.

(a) What is the dual origin of the prostate?

(b) What are the origins of the bulbourethral gland?

151 (a) Which embryonic germ layer lines the secondary yolk sac?

(b) Give two derivatives of the yolk sac.

(c) What is its function in Week 2-3?

151

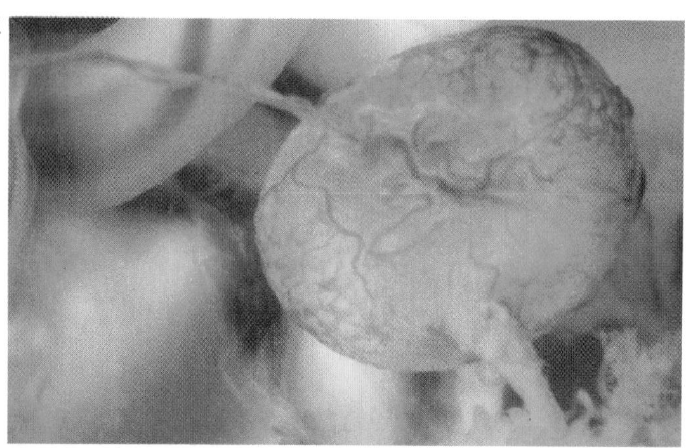

152 The thorax and abdomen of a 20mm CR (Stage 19) embryo dissected to show the liver.
(a) What is the relative size of the liver in Week 6?
(b) Which activity is responsible for the relatively large liver size in the second month?

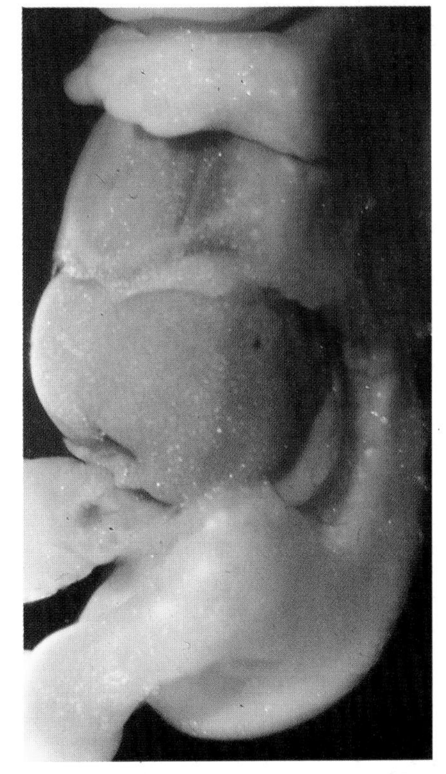

152

153 The pancreas removed and viewed from behind in a 48mm CR fetus.
(a) How does the pancreas form?
(b) When does insulin formation begin?
(c) An annular pancreas may cause the obstruction of which part of the gut shortly after birth?

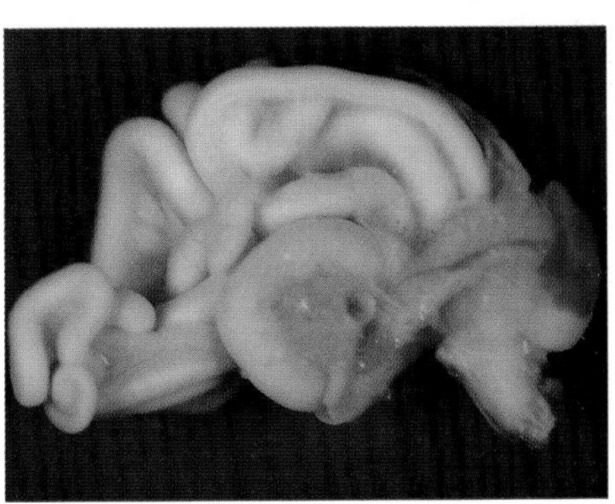

153

154

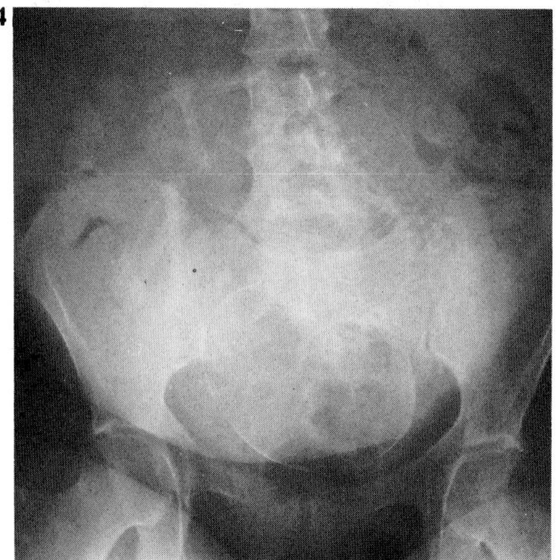

154 An X-ray of the abdomen and pelvis in pregnancy.
(a) Fetal movements detected by the mother are known by what name?
(b) When can limb movements first be detected by ultrasound?
(c) Where do the ribs ossify first?

155 (a) List some of the bones of the vault of the skull which ossify in membrane.
(b) How do the flat bones of the skull grow to accommodate the growth of the brain?

155

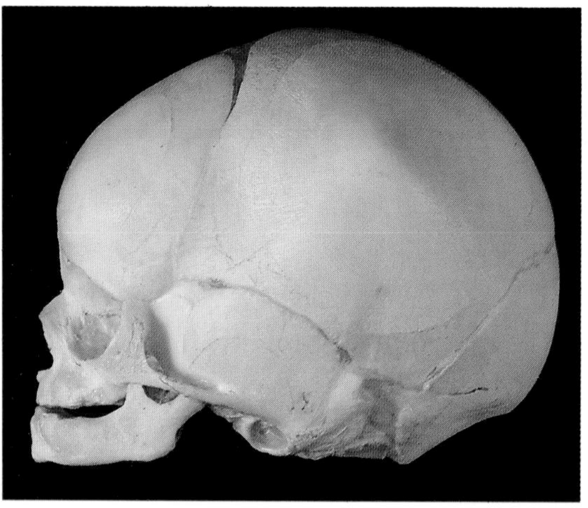

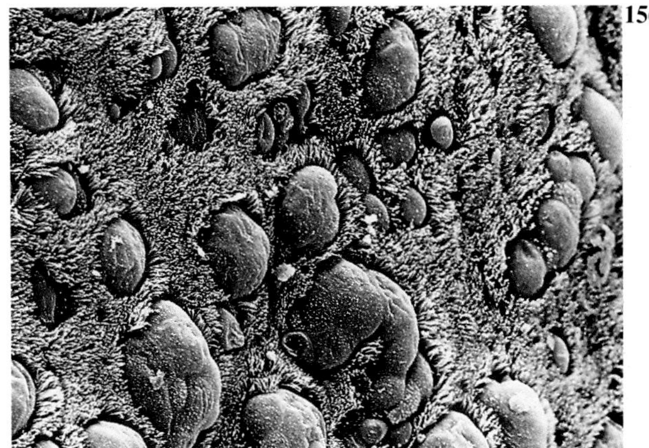

156 A scanning electron micrograph of fetal oesophagus.
(a) Is this appearance normal?
(b) When is the oesophageal lumen almost obliterated by the proliferation of the tube endoderm?
(c) What are the origins of the striated and non-striated muscles of the oesophagus?

157 A sagittal section through the oral cavity of a 125mm CR fetus.
(a) When are the lips (labia) and gums (gingiva) separated into two components?
(b) How are the teeth separated from the lips and cheeks?

157

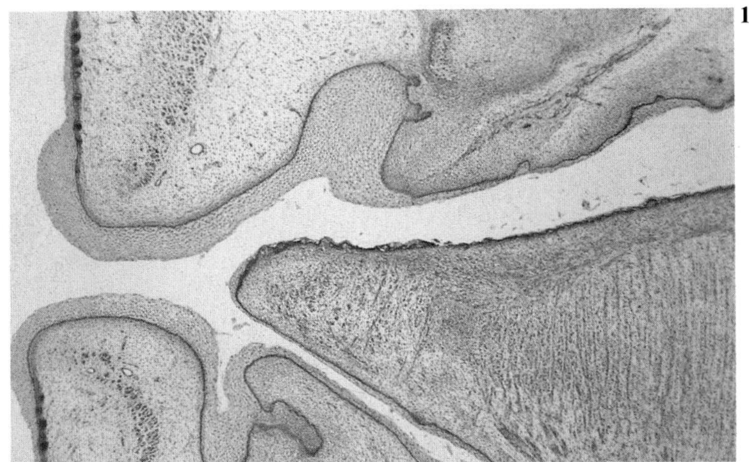

158

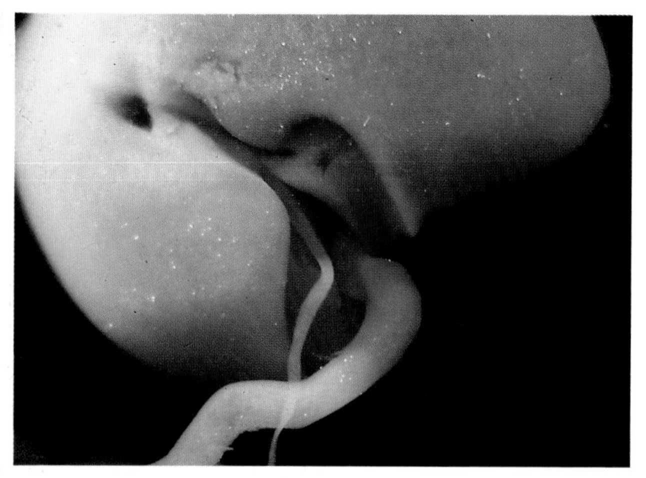

158 Early liver with the falciform ligament removed.
(a) Identify the blood vessel.
(b) What remnant of this blood vessel is present in the adult?
(c) What happens to the right umbilical vein?

159 A longitudinal section through the eye of a 10mm CR embryo.
(a) What are the optic or choroidal fissures?
(b) How are the blood vessels incorporated into the optic stalk?
(c) Identify the vessels.

159

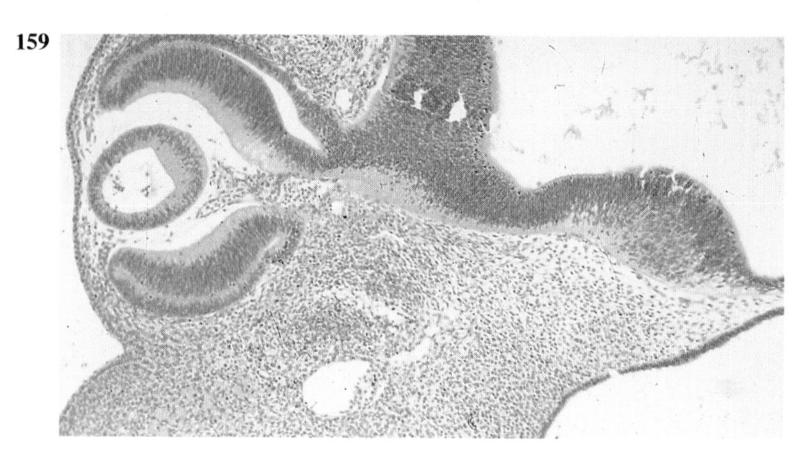

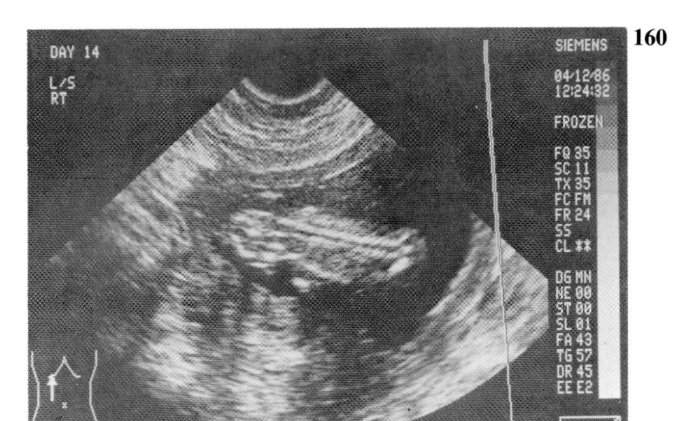

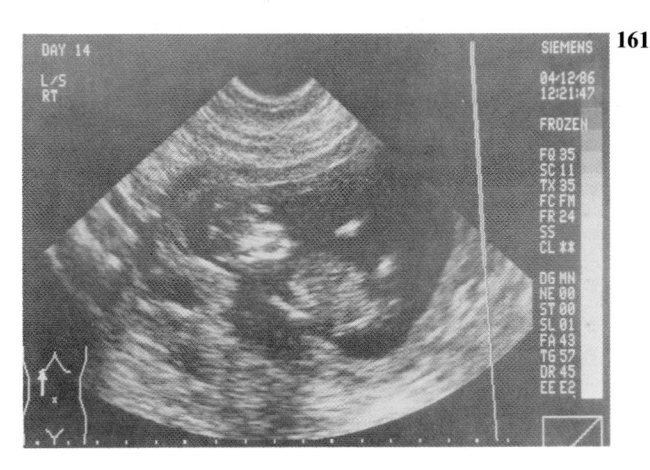

160, 161 Ultrasound scans: longitudinal views.
(a) What is anencephaly?
(b) Can this be diagnosed by ultrasound?
(c) With which cranial abnormality is it always associated?

162

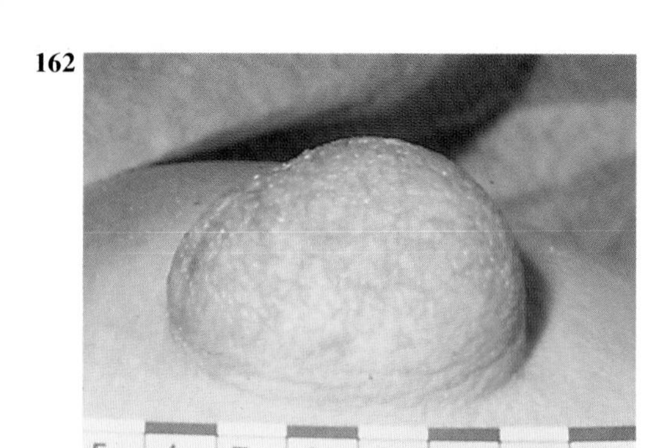

162 (a) How can an umbilical hernia be distinguished from failure of the intestine to return to the abdomen during reduction?
(b) What is the term for failure of reduction?

163

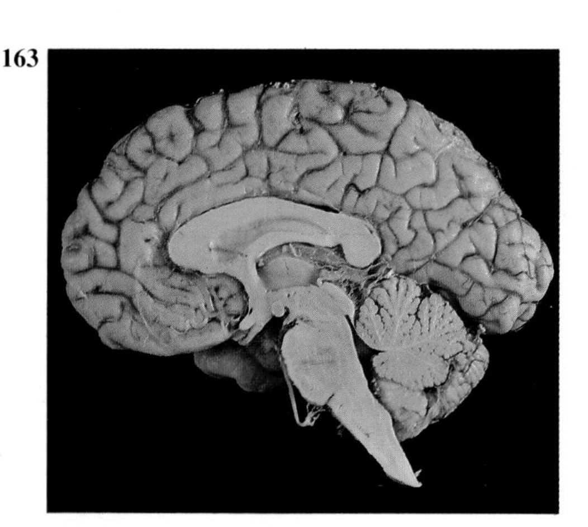

163 A sagittal section of adult brain.
(a) How does it become C-shaped?
(b) How does the falx cerebri form?

164

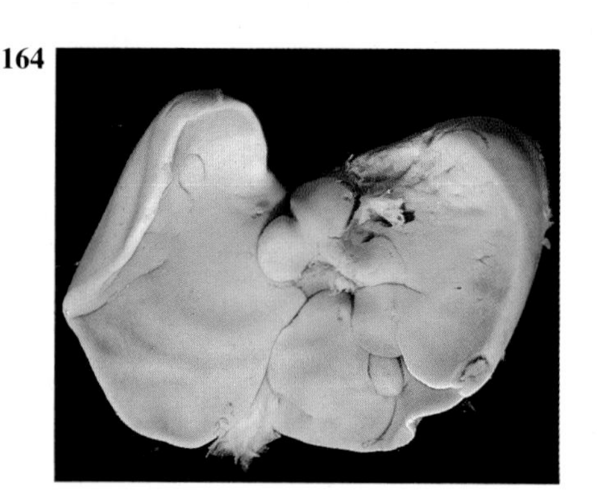

164 The liver viewed from below.
(a) The caudate and quadrate lobes form from which lobe?
(b) Are congenital anomalies of the liver common or rare?
(c) The ventral mesentery gives rise to what three structures?

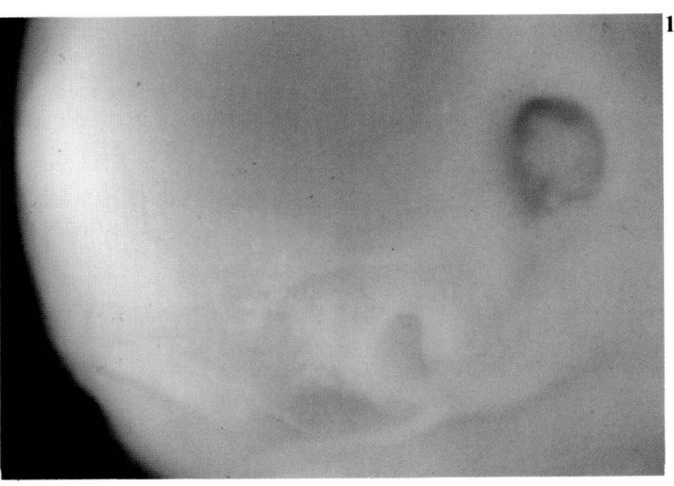

165 The face of an early embryo.
(a) What will the frontonasal prominence form?
(b) Each nasal pit leads into which structure?
(c) How do the primitive choanae (nares) form?

166 A transverse section of the brain in a 10mm CR embryo.
(a) What are the three layers of the early brain and spinal cord?
(b) What additional layer is added to the brain and how is it added?

166

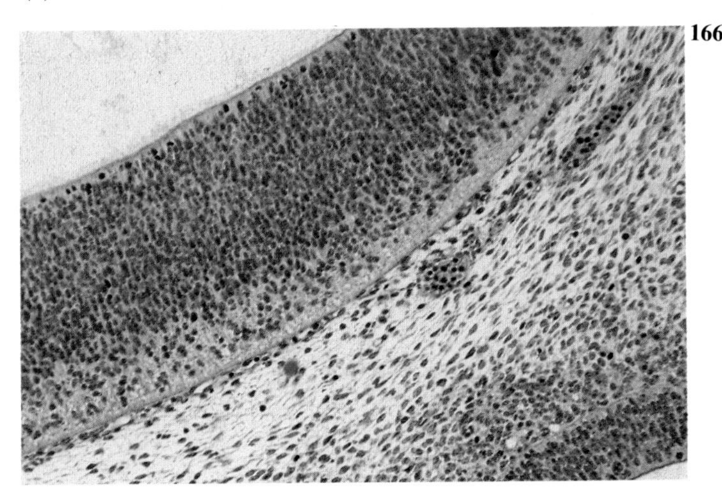

167

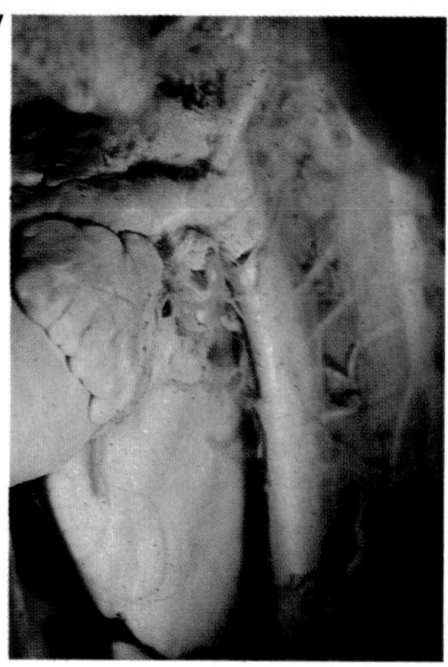

167 The heart and great vessels viewed from the left.
(a) Identify the vessel connecting the pulmonary trunk with the aorta.
(b) What happens to this vessel at birth?
(c) Which substance released by the lungs mediates the change at birth?

168 A transverse section illustrating the pericardial cavity.
(a) By which mesentery is the heart initially suspended from the dorsal wall of the pericardial cavity?
(b) What is the opening through which the right and left sides of the pericardial cavity communicate?
(c) How does this opening form?

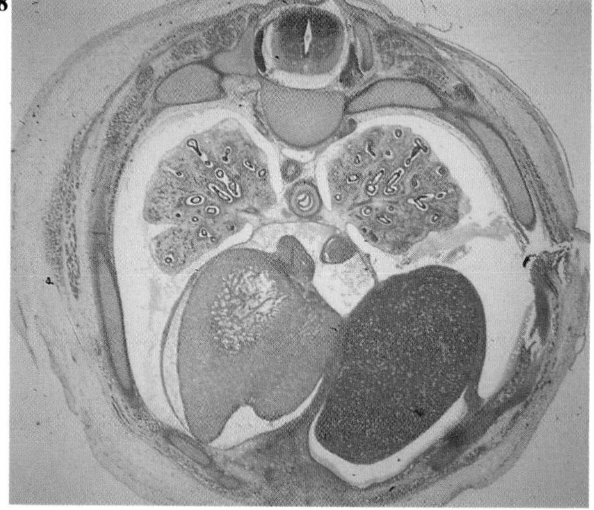

169

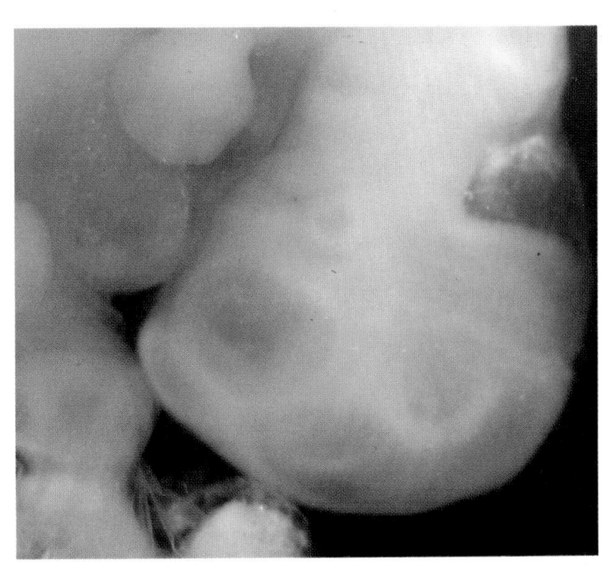

169 The brain viewed from the side (14mm CR, Stage 17).
(a) What are the three major regions of the early brain?
(b) Which two vesicles grow out from the telencephalon?
(c) Which adult feature of the brain marks the site of the closed rostral neuropore?

170

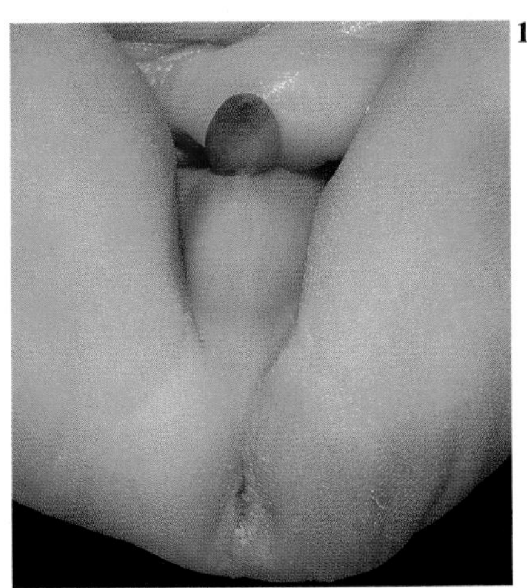

170 (a) The skin of the anal region is continuous with the epithelium of the anal canal. What happens to the epithelium histologically at the anus?
(b) About 2cm superior to the anus is the anocutaneous line (White Line of Hilton). What histological changes occur here?
(c) The pectinate line is the approximate former site of which embryonic membrane?

171

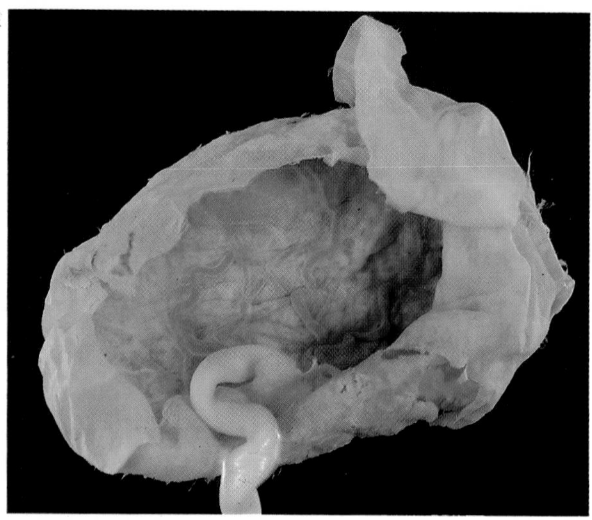

171 (a) Is this the maternal or fetal aspect of the placenta and how can you differentiate between them?
(b) Give three functions of the placenta.

172 A tooth in the bell stage of development.
(a) Where is enamel laid down first?
(b) The dental papillary mesoderm will differentiate into what tissue?
(c) What are Tomes' dentinal fibres?

172

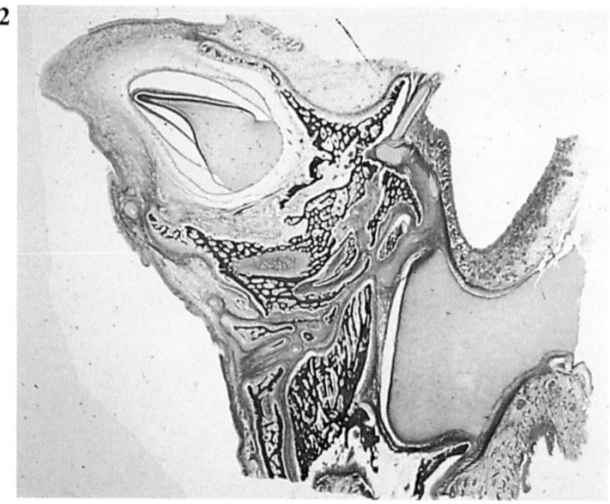

173 A coronal section of the head (32mm CR fetus).
(a) Identify the three palatal processes.
(b) When do the palatal processes elevate and fuse?
(c) In what order do the processes fuse?

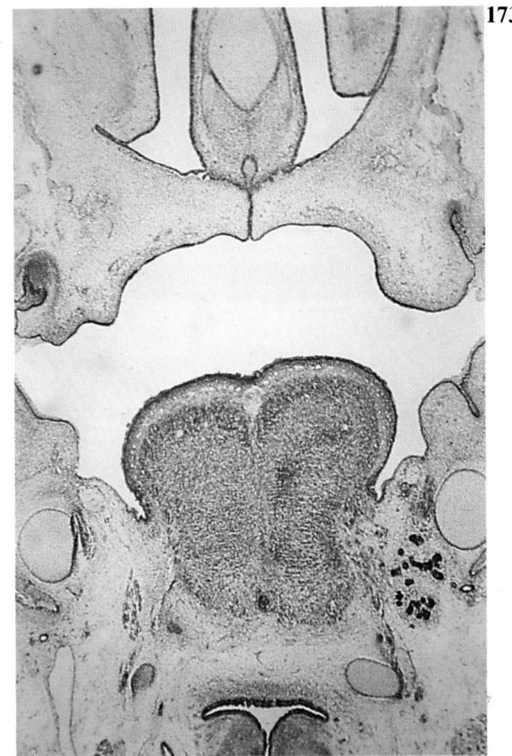
173

174 The parathyroid gland (*arrow*) and thyroid tissue of a fetus.
(a) Where do the parathyroid glands develop?
(b) How do the parathyroids assume their adult position?
(c) When is parathyroid hormone produced?

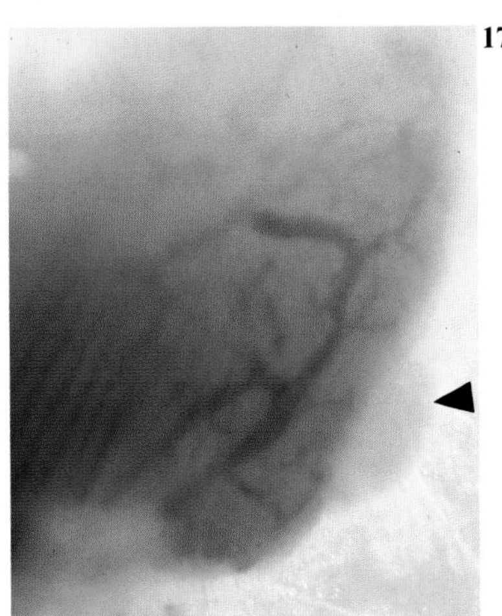
174

175

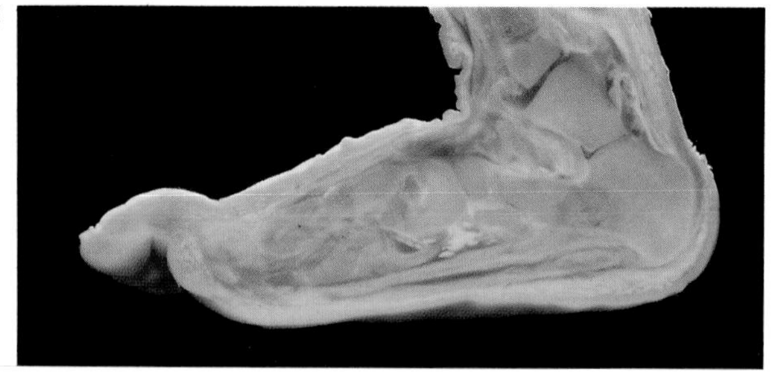

175 A longitudinal section of the foot (220mm CR fetus) showing the os calcis.
(a) When does this centre of ossification normally appear?
(b) How is this ossification centre used in a post-mortem examination?
(c) Does this bone have red marrow at birth?

176

176 The fetal skull and maternal pelvis positioned as in child-birth.
(a) What is restitution?
(b) What are the stages of labour and how many are there?

177

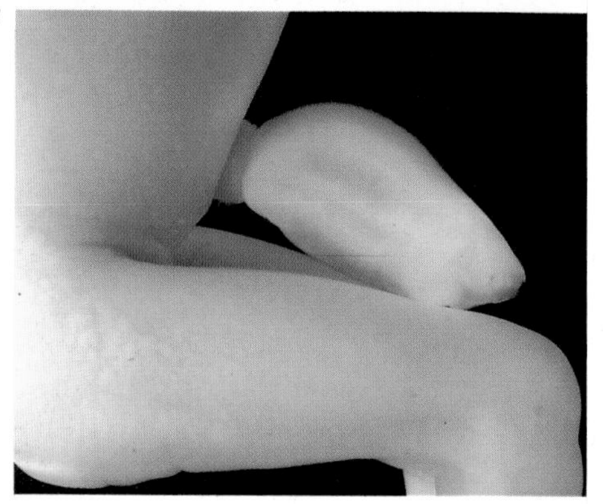

177 A fetus viewed from the right side.
(a) When does the storage of subcutaneous fat increase dramatically and the body become plump?
(b) What are the two types of fat?
(c) Where is brown fat found and what is its importance?

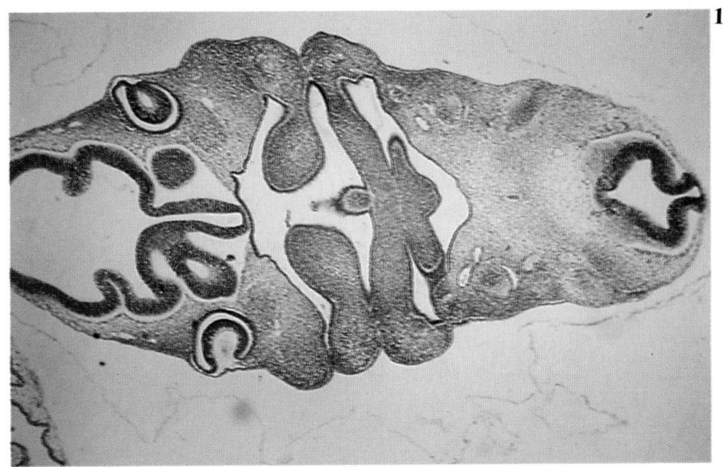

178 A transverse section of the head in a 7mm CR embryo.
(a) The pharyngeal pouches are lined with what type of tissue?
(b) What is the structure called where the foregut epithelium contacts the ectoderm of the branchial groove?
(c) Does the fifth pharyngeal pouch exist?

179 A scanning micrograph of fetal intestine.
(a) When are villi present?
(b) When do the intestinal enzymes first appear?
(c) Describe the muscle layers at birth.

180

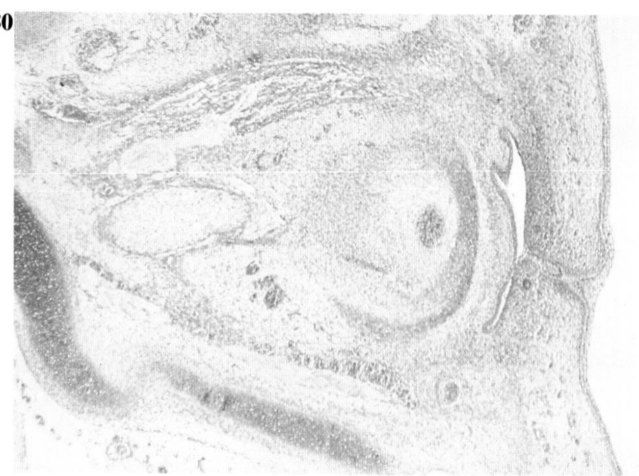

180 A sagittal section of the eye.
(a) The early lens is a hollow vesicle. What are its derivatives?
(b) What is the fate of the hyaloid artery and vein?

181 The heart viewed from behind. The left ventricle has been displayed to show a hole-in-the-heart.
(a) What is the detection rate for isolated ventricular septal defects?
(b) In which syndrome is an endocardial cushion defect (atrioventricularis communis) common?
(c) What is the frequency in this syndrome?

181

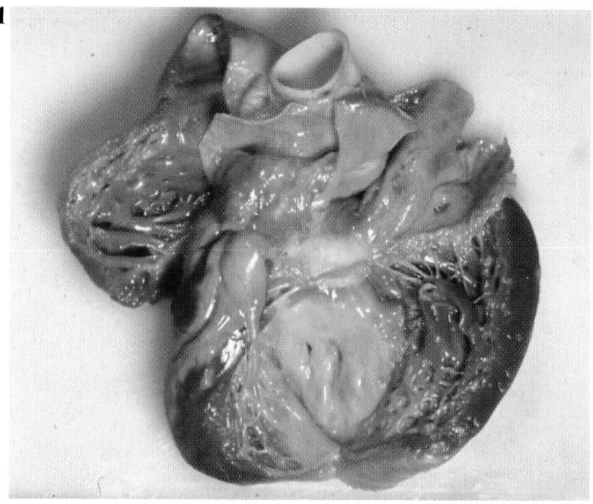

182

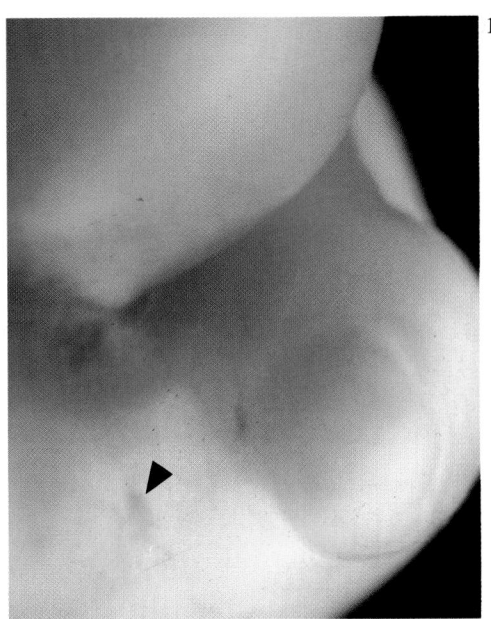

182 The proctodeum (*arrow*) of an embryo viewed from the front.
(a) How does the proctodeum (or anal pit) form?
(b) Which structure forms where the urorectal septum fuses with the cloacal membrane?
(c) At birth the rectum is usually distended by its contents. What are they?

183 A horizontal section through the branchial region of a 4mm CR (Stage 11) embryo.
(a) From which pouch does the palatine tonsil form?
(b) From which pouch does the thymus gland form?
(c) What is the fate of the ventral first and second pouches?

183

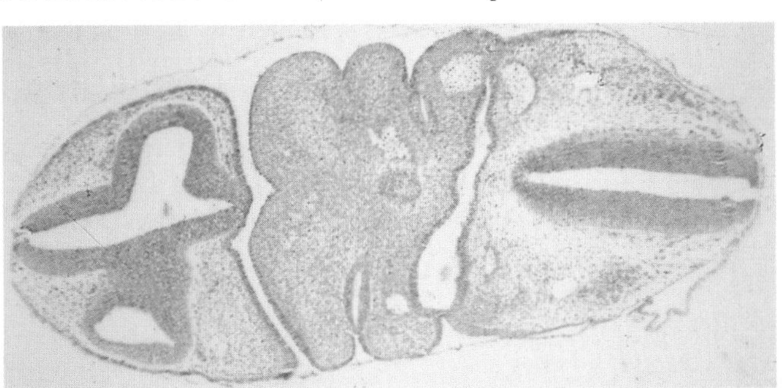

184

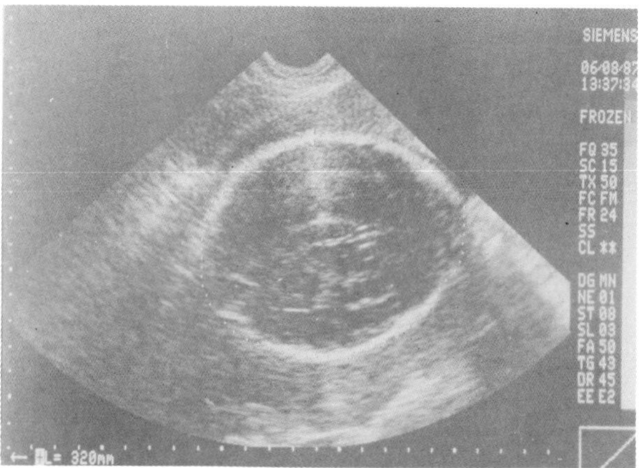

185

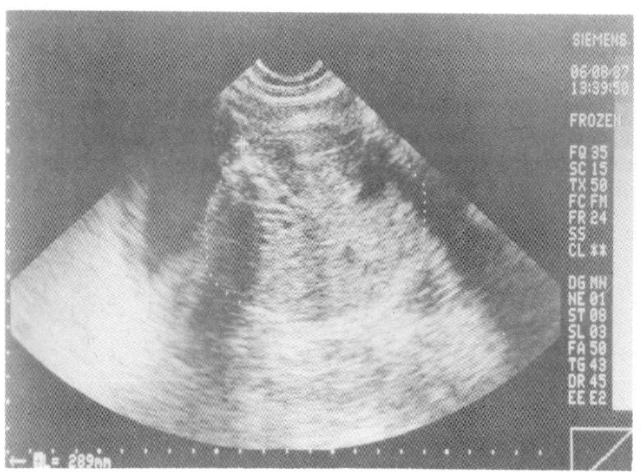

184, 185 An ultrasound scan of a fetus of 33 weeks gestation.
(a) If the femoral length is small compared to the biparietal diameter, what anomaly must be excluded?
(b) In multiple pregnancies which femoral length should be used to determine the gestational age?
(c) In studies of fetal growth (as opposed to fetal age) what two measurements are made?
(d) What effect can maternal smoking have on fetal development?

186 (a) What is the buccal fat pad?
(b) What is believed to be its function?
(c) Why does a fetus born at Week 22 look old and wizened?

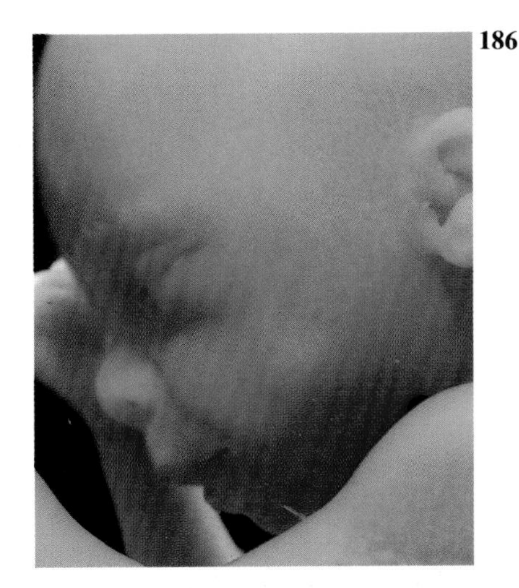
186

187 The hindbrain of a 14mm CR (Stage 17) embryo viewed from above and behind.
(a) From which part does the cerebellum form?
(b) From which part does the pons form?
(c) What is the relationship of the otic placode to the hindbrain?

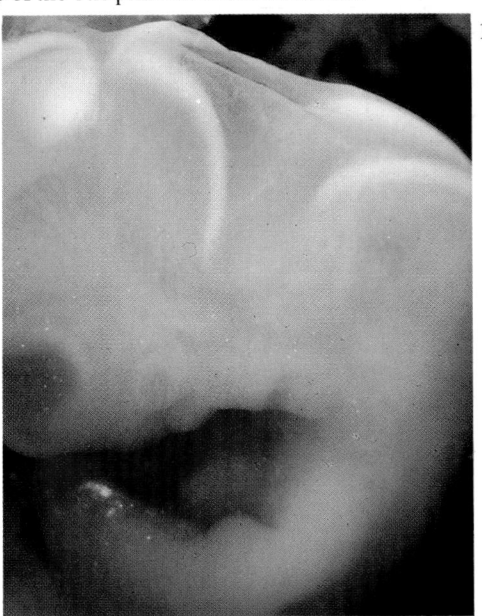
187

188

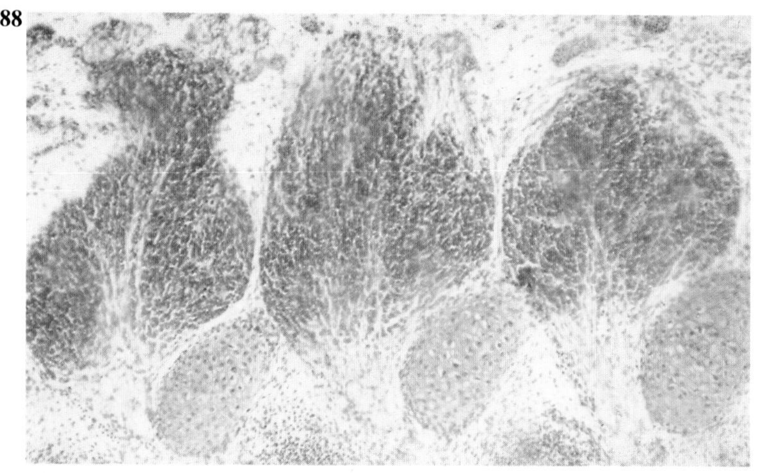

188 A longitudinal section of the back of a 20mm CR embryo.
(a) Unipolar neurons in the spinal ganglia arise from which embryonic cells?
(b) The central processes of these neurons enter the spinal cord as which structures?

189 (a) Where does the genital gland appear in relation to the mesonephroi?
(b) What is the fate of the cranial mesonephroi?
(c) What is the fate of the mesonephric duct in the male?

189

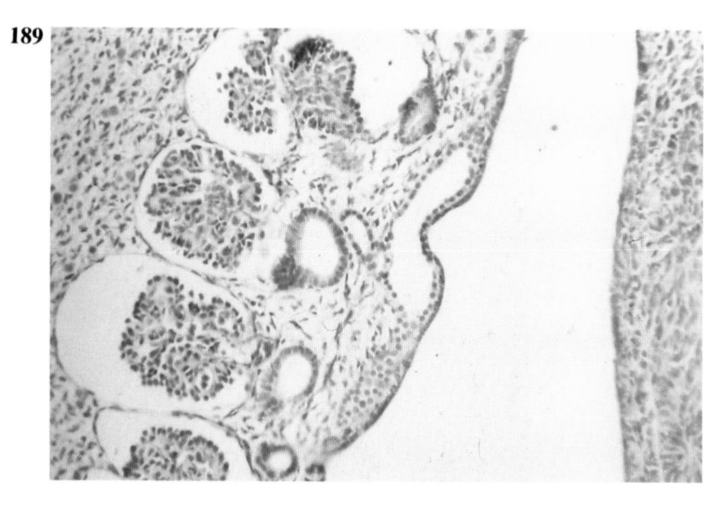

190 The nose and mouth of a 51mm CR fetus.
(a) What structure is formed by the fusion of the medial nasal prominences with the maxillary prominences?
(b) What does this structure give rise to?
(c) The lateral nasal prominences form which adult structure?

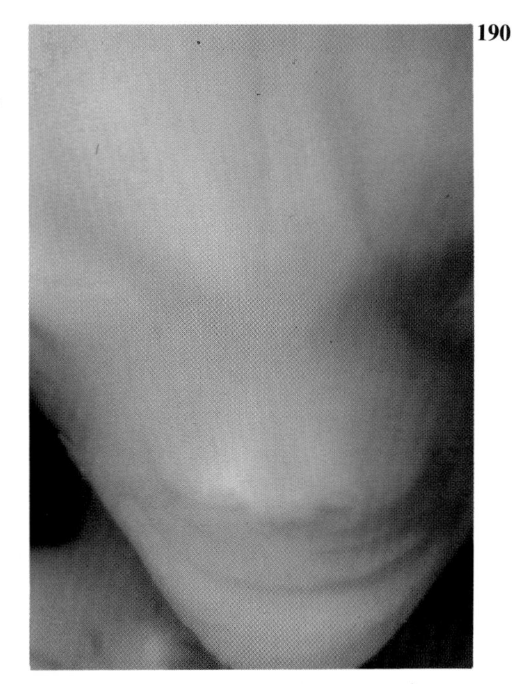
190

191 The suprarenal (adrenal) glands and kidneys viewed from behind.
(a) Which of the two parts of the suprarenal gland is responsible for its comparatively large size?
(b) Which cells form the chromaffin cells?
(c) Give three substances produced by the cortex and medulla.

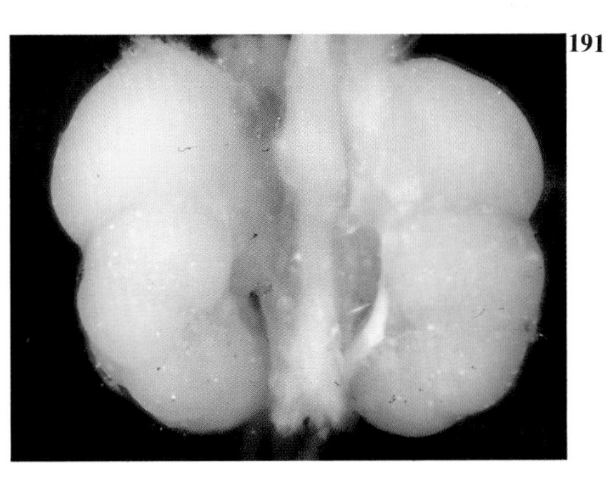
191

192

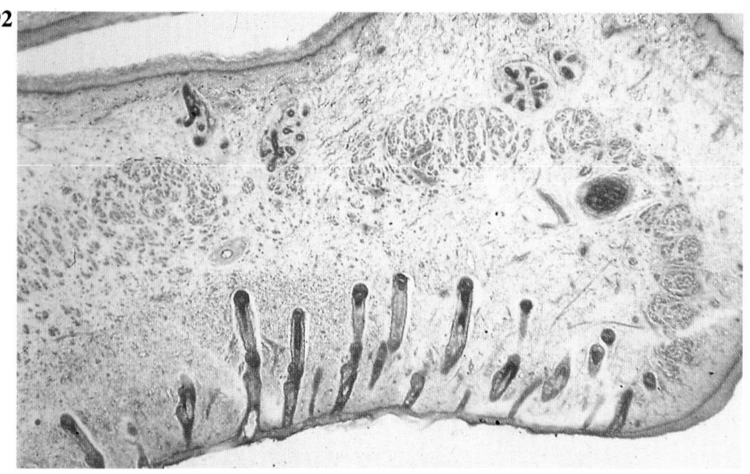

192 A transverse section of fetal lower lip.
(a) Which primary embryonic layer contributes to the lining of the mouth?
(b) What membrane separates the two primary embryonic layers in the early embryo?
(c) Which two primitive cavities are brought into communication when this membrane ruptures at Day 24?

193

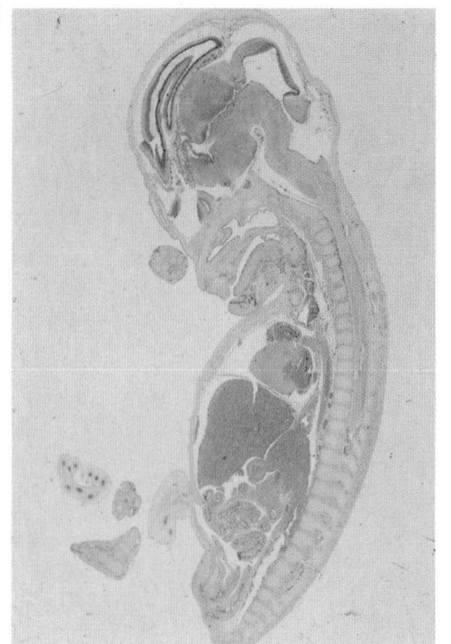

193 A longitudinal section of a 40mm CR fetus to illustrate the anal canal.
What is the blood supply to the upper two-thirds and to the lower one-third of the anal canal?

194

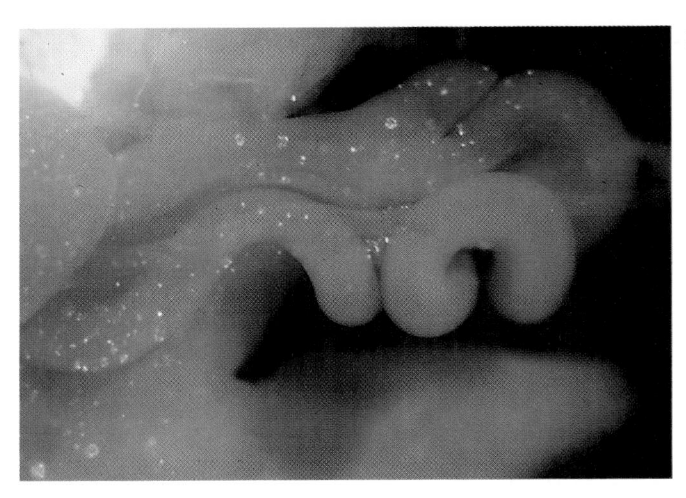

194 The midgut herniation of a 20mm CR embryo. The umbilical cord has been dissected off.
(a) Which artery supplants the early vitelline arteries?
(b) Identify two landmarks present on the gut.
(c) Why does the gut herniate into the umbilical cord in Week 6?

195

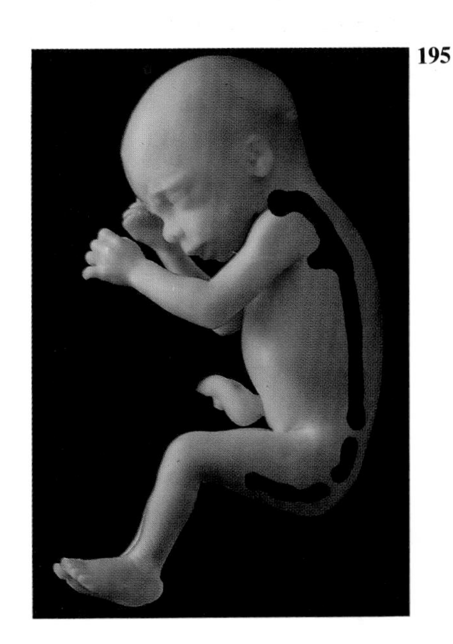

195 (a) How many lymph sacs form initially?
(b) How do the lymphatics spread throughout the body?
(c) What will mesenchyme surrounding the sacs form?

196

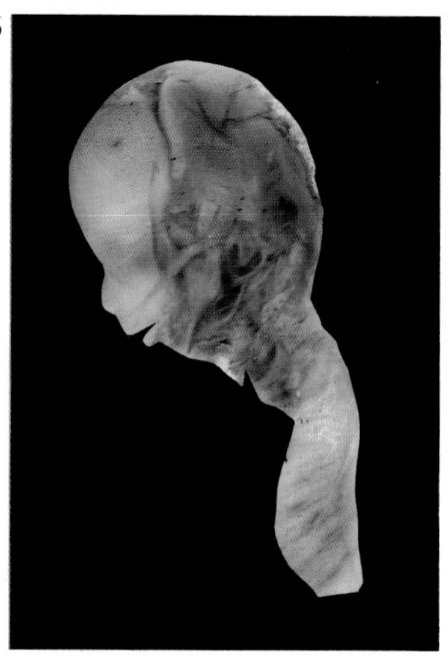

196 (a) What are the nerves of the first, second and third branchial arches?
(b) What ligaments are derived from the first arch?
(c) Name some of the muscles derived from the second branchial arch.

197 A transverse section of the heart and lungs.
(a) Which structures direct blood from the inferior vena cava through the foramen ovale?
(b) Which two embryonic vessels contribute to the superior vena cava?
(c) The foramen ovale is an opening in which embryonic septum?

197

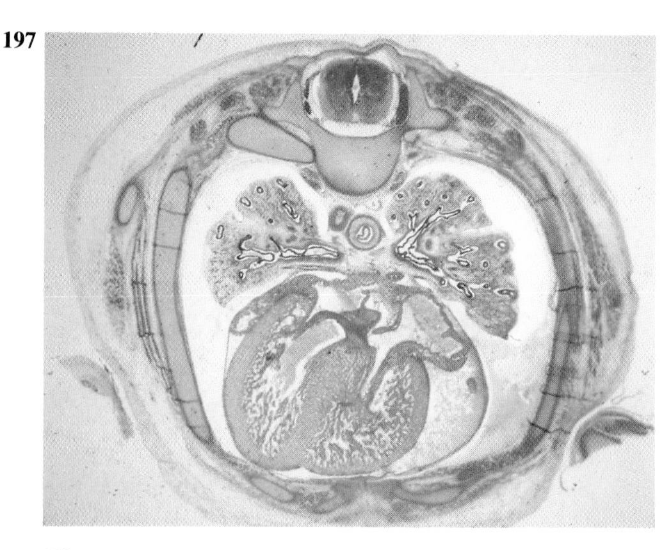

198

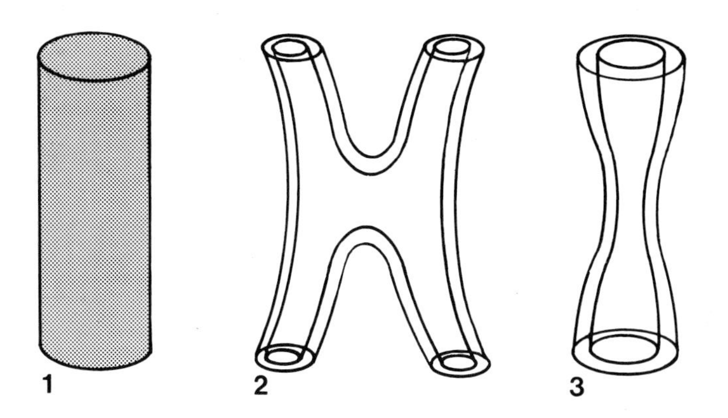

198 (a) Define fistula, atresia and stenosis. Identify the relevant drawings.

199

199 A 14mm CR (Stage 17) embryo viewed by transmitted light. The midbrain is most prominent at the midbrain flexure.
(a) What forms the lateral walls?
(b) Name five structures which reduce the wide lumen of the midbrain to the narrow cerebral aqueduct.

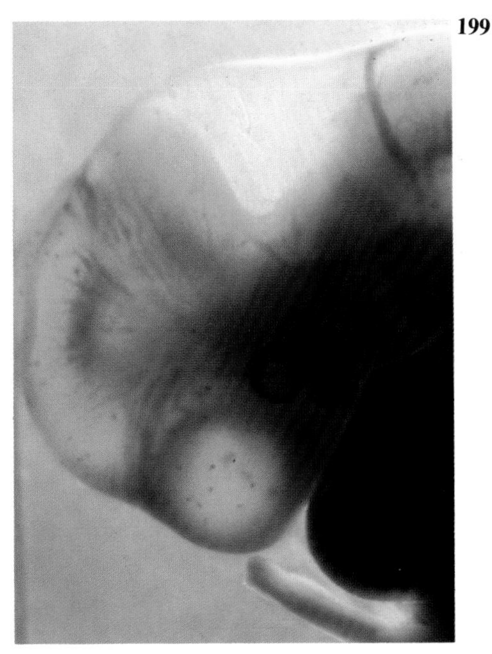

200

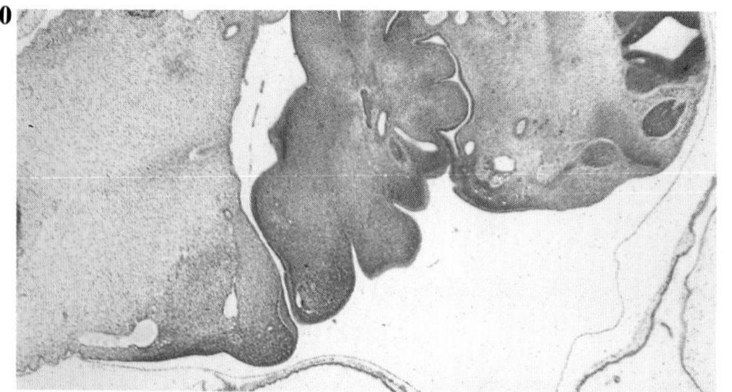

200 A transverse section through the head region of a 9mm CR embryo.
(a) How many pairs of branchial arches are there in an embryo?
(b) What adjacent region do they support?
(c) What is the ectodermal depression formed as the second branchial arch overgrows the third and fourth arches?

201

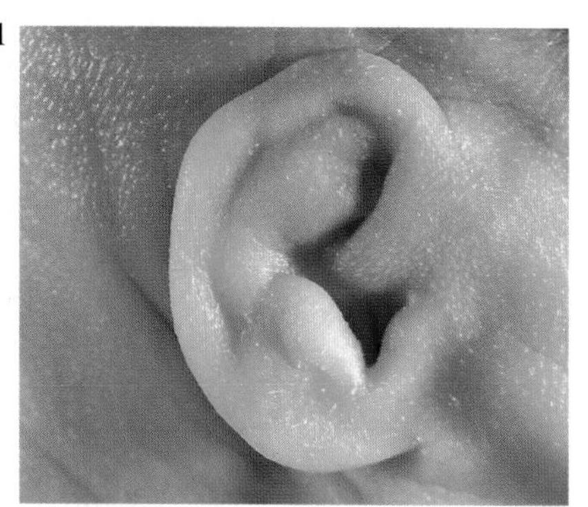

201 (a) The external acoustic meatus is a derivative of which structure?
(b) What forms the early tympanic membrane?
(c) What tissues contribute to the fetal tympanic membrane?

202

202 Fresh lung tissue from a 220mm CR infant has been placed in a beaker of water.
(a) What can be deduced about the infant?
(b) What are the constituents of the fluid filling the lungs at birth?
(c) Give three ways in which fluid is removed from the lungs at birth.
(d) As the lungs expand with their first breath, what occurs in the heart?

203

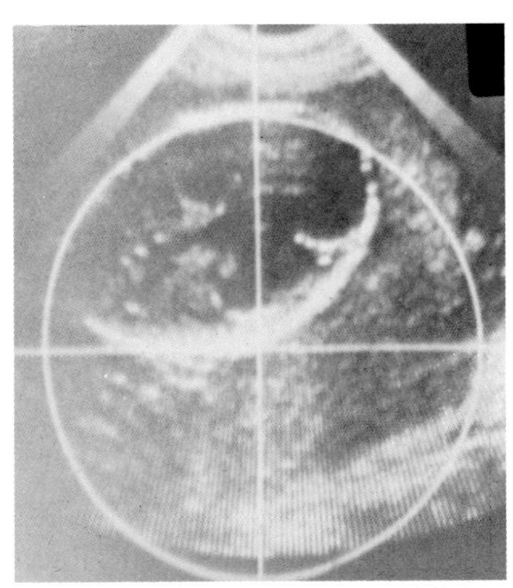

204

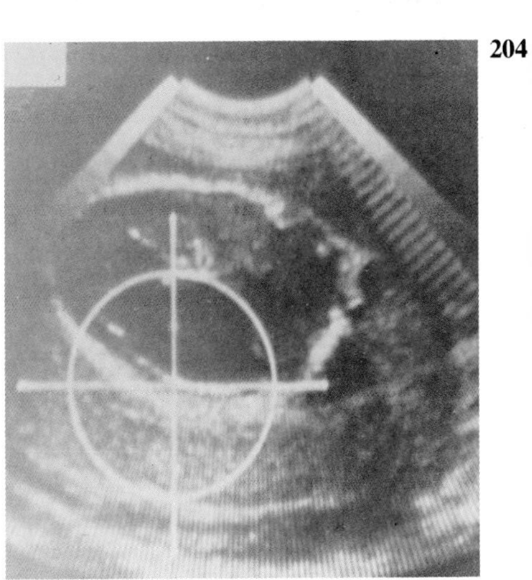

203, 204 An ultrasound scan showing the fetal head. The median and lateral foramina form independently of the pressure of the cerebrospinal fluid (C.S.F.) in the fourth ventricle.

(a) If the median and lateral foramina are blocked and C.S.F. cannot escape, what condition results?

(b) How can this condition be diagnosed?

205

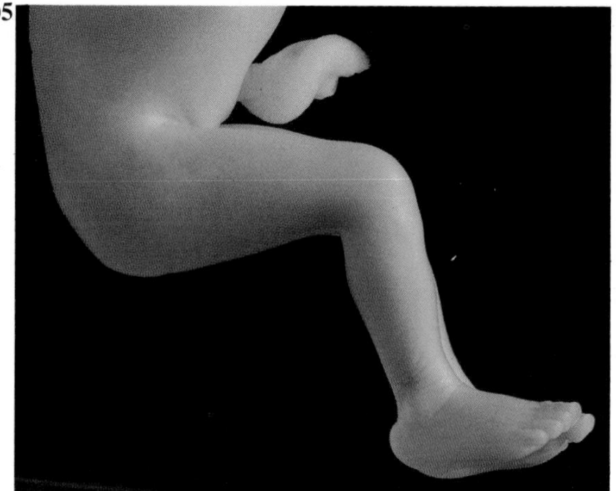

206

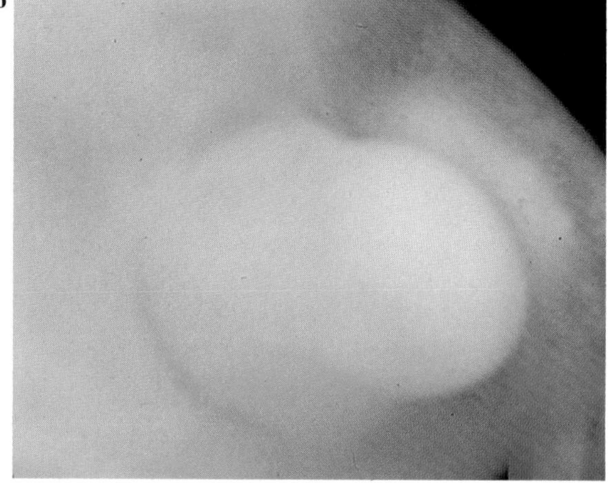

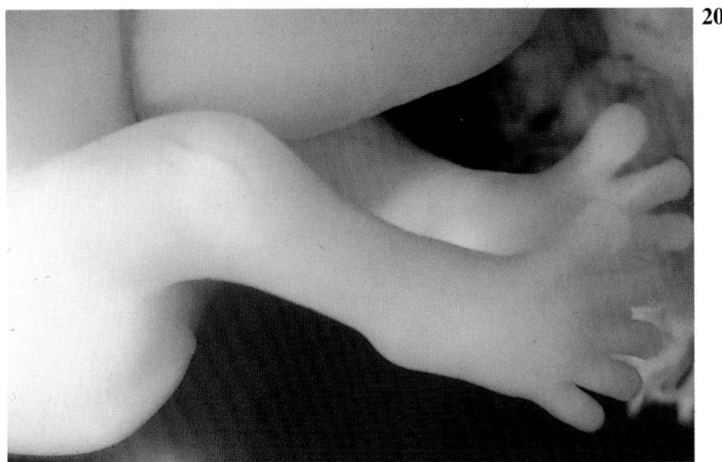

208

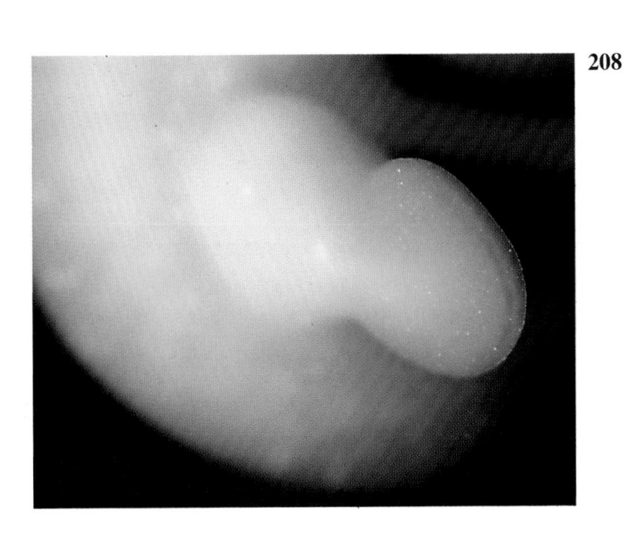

205-208 Rotation of the lower limb.
(a) What is the correct developmental sequence of these pictures?
(b) When does rotation of the knee occur?
(c) What position does the neonatal lower limb assume?

ANSWERS

1 (a) Posterior aspect of the upper part of the uterus.
(b) One which occurs outside the uterus (e.g. uterine tubes, abdominal cavity). Over 90 per cent of ectopic pregnancies occur in the uterine tubes.
(c) If implantation occurs in the lower uterus, the placenta may cover the internal os. The fetus would be obstructed during parturition.

2 (a) Two arteries; one vein.
(b) Umbilical vein.
(c) Proximally as internal iliac arteries and branching superior vesical arteries; distally as the medial umbilical ligaments within the medial umbilical folds to the umbilicus.
(d) Irregular projections of blood vessels.

3 (a) Ectoderm externally, foregut endoderm internally with a core of mesenchyme between them. Neural crest cells migrate into each arch and surround the core of mesenchyme. An artery, a cartilaginous bar and a nerve will be present.
(b) Muscles, skeletal and connective tissues.
(c) Branchial arch I trigeminal; branchial arch II facial; branchial arch III glossopharyngeal; branchial arch IV and VI superior laryngeal of vagus; recurrent laryngeal of vagus.

4 (a) Lateral ventricle.
(b) Interventricular foramen.
(c) Third ventricle.

5 (a) Posterior body fold or tail fold.
(b) Regression, cell death and rapid increase in size of the buttock region.
(c) Coccyx.

6 (a) Syndactyly: webbing, due to a lack of differentiation between two or more digits.
(b) Meromelia: partial absence of a limb(s); amelia: complete absence of a limb(s).
(c) Minor limb defects are relatively common.

7 (a) Glomeruli.
(b) Yes, the fetal kidney is normally lobulated.
(c) About six weeks after birth.

8 (a) Day 28.
(b) Endoderm of the respiratory diverticulum (the laryngotracheal tube) of the foregut and splanchnic mesenchyme.

9 (a) Primitive pulmonary vein and entry of the four pulmonary veins.
(b) Right atrium and a small venous return from the pulmonary veins.
(c) In the left atrium, forcing the membranous septum primum against the muscular septum secundum to close the foramen ovale.

10 (a) They fuse only at the fourchette.
(b) Labia minora.

(c) They fuse cephalically to form the mons pubis, and caudally to form the posterior labial commissure. The remaining unfused swellings form the labia majora.

11 (a) As a result of changes in relative position and proportion, the eyes and nostrils move medially on the face.
(b) Frontonasal prominence.
(c) 5-8 weeks.

12 (a) Initially membranous, in later growth cartilages develop at both ends.
(b) Cartilaginous.
(c) Female.

13 (a) Failure of fusion of the mesenchymal masses of the two lateral palatine processes, with the nasal septum, with the primary palate or median palatine process.
(b) The maxillary division of the trigeminal grows from the maxillary prominences into the segments.
(c) The ophthalmic division enters from above.
(d) Clefts involving the upper lip occur approximately 1 in 1000 births; more males are affected than females.

14 (a) Anterior fontanelle.
(b) The process begins three months after birth and is usually completed by 18 months-2 years.
(c) It is obliterated by the ingrowing of the borders of the membrane bone.

15 (a) Mesenchyme surrounding the neural tube condenses to form the primitive meninx.
(b) Dura mater.
(c) Neural crest may contribute to the pia-arachnoid.

16 (a) Cranial nerve I.
(b) Week 5-6.
(c) Special sensory; outgrowths of the brain I and II (and VIII). Branchial arch nerves: V, VII, IX, X (and XI). Somatic efferent: IV, VI, XII, and the greater part of III. Somatic efferents are homologous with the ventral roots of spinal nerves.

17 (a) Floor of the myelencephalon.
(b) Sulcus limitans.
(c) Corticospinal fibres grow down from the telencephalon.

18 (a) Mesonephroi.
(b) Nephrogenic cord; lower cervical and upper thoracic intermediate mesenchyme.
(c) Cloaca.

19 (a) Primitive ventricles.
(b) Mesenchyme.
(c) The muscular portion derives from subendocardial tissue from the right and left bulbar ridges, and the fused atrioventricular endocardial cushions. The membranous portion derives from tissue from the right side of the fused endocardial cushions, then fuses with the muscular portion and the aorticopulmonary septum.

20 (a) As a downgrowth of thickened ectoderm in the pectoral region. Secondary epithelial cords grow into underlying mesenchyme and then become canalised (Week 32-36).
(b) At full term or in the neonate, a proliferation of mesenchyme below the areola elevates it above the adjacent skin.

21 (a) The flexor compartment by the anterior divisions of the trunks of the brachial plexus and the extensor by the posterior divisions.
(b) The preaxial forms the radial border and is marked by the position of the cephalic vein. The postaxial forms the ulnar border and is marked by the position of the basilic vein.
(c) *In situ*.

22 (a) Muscles are primarily from cervical myotome 4 with contributions from 3 and 5, which invade the septum transversum. Mesenchyme of the septum then migrates into the other parts of the fused diaphragm and differentiates into myoblasts and then muscle.
(b) Posterolateral defect of the diaphragm.
(c) Herniation of abdominal viscera into the thoracic cavity.

23 (a) Prosencephalon: forebrain.
(b) They are outgrowths of the brain.
(c) Retina, ciliary body, iris and dilator and sphincter pupillae muscles of the iris.
(d) Mainly in the first three weeks after birth, completed by ten weeks.

24 (a) Relatively well-developed when compared with the thin muscular layer of the rectum.
(b) The upper two-thirds is from hindgut; the lower third is from proctodeum.

25 (a) From mesenchyme in the dorsal mesogastrium.
(b) Megakaryocytes, myeloblasts, erythroblasts.
(c) In the greater omentum.
(d) Functioning thymus.

26 (a) Spina bifida.
(b) Alphafetoprotein.
(c) Yes, the fetus can empty its bladder and kick. The nerves only become damaged when they are no longer supported by amniotic fluid.

27 (a) The urethra opens on the dorsal surface of the penis.
(b) Failure or incomplete fusion of the urogenital folds which results in an incomplete penile urethra.
(c) Androgens from the fetal testes.

28 (a) Small and large intestine.
(b) Growth of the terminal portion of the caecum lags behind the remaining caecum. A long blind sac is formed.
(c) As the colon elongates, they descend so that the appendix assumes a retrocaecal, retrocolic, or pelvic position.
(d) No. The caecum tapers into the appendix.

29 (a) Endodermal lining of the cranial laryngotracheal tube (respiratory diverticulum) and mesenchyme from the fourth and sixth branchial arches.

(b) Epiglottal swelling from the third and fourth branchial arches, and two arytenoid swellings.
(c) The fourth and sixth branchial arches.

30 (a) Sinus venosus.
(b) Umbilical veins, vitelline veins and common cardinal veins.
(c) They are fixed by the branchial arches and septum transversum.

31 (a) Epithalamus, thalamus and hypothalamus.
(b) Roof and dorsal portion of the lateral wall of the diencephalon.
(c) In 80 per cent of brains, the two thalami fuse and form a bridge of grey matter across the third ventricle.

32 (a) As the sac grows, the blood vessels in this area are compressed and degenerate. The sac becomes smooth in this region (chorion laeve).
(b) Amnion.

33 (a) First or mandibular and second or hyoid.
(b) Mandibular and maxillary.
(c) Upper part of body of hyoid bone and lesser cornu, styloid process, stapes, stylohyoid ligament, facial nerve (VII), muscles of facial expression, stapedius, stylohyoid, and posterior belly of digastric.

34 (a) Metaphysis.
(b) From 12 weeks to term.
(c) Biparietal diameter.

35 (a) Testis.
(b) Week 7.
(c) An outer cortex and inner medulla.
(d) The cords are prominent in the male but break up into cell clusters in the female.

36 (a) Polycystic lungs.
(b) Dilation of terminal bronchi.
(c) Kidneys, liver.

37 (a) Inferior part of the head and uncinate process.
(b) Part of the head, neck, body and tail.
(c) The main pancreatic duct forms from the duct of the ventral bud and a distal portion of the dorsal bud duct. The accessory duct forms from the proximal part of the dorsal duct.

38 (a) Adenohypophysis (outpouching of the stomodeum) and neurohypophysis (diverticulum of the diencephalon).
(b) Pars anterior, tuberalis, and intermedia.
(c) Pars nervosa, the infundibular stem and the median eminence.

39 (a) The lips fuse at the angles.
(b) Second branchial arch mesenchyme.
(c) First branchial arch mesenchyme.
(d) Yes, the facies are affected and are one clinical manifestation of the fetal alcohol syndrome; others include pre- and postnatal growth deficiencies, joint or congenital heart abnormalities, mental retardation, etc.

40 (a) The choroid plexus produces cerebrospinal fluid (CSF).
(b) Lateral, third and fourth ventricles.
(c) Pia mater and blood vessels invaginate the thin inner ependymal walls of the hemispheres, the roof of the third ventricle and fourth ventricle.

41 (a) Week 6-10.
(b) A and B at 5-6 weeks.
(c) Smooth muscle and connective tissue of the blood vessel.

42 (a) Two umbilical arteries, one umbilical vein,. allantois, coelom, and primary mesoderm cells.
(b) Wharton's jelly.
(c) Growth of the amnion forces the yolk stalk against the embryonic stalk and then covers the entire contents with a sleeve of amnion.

43 (a) It becomes the urachus and during Week 5-8 the extra-embryonic part degenerates.
(b) Mesonephric ducts and ureters.
(c) Abdominal.

44 (a) The mandibular prominences fuse first.
(b) Two halves fuse at the midline suture (symphysis menti) at the beginning of the second year.
(c) Meckel's cartilage.
(d) Very broad, 140°.

45 (a) Endoderm (foregut).
(b) Mesenchyme.
(c) Apparent ascent of the pharynx.

46 (a) Digital (radial) ridges or rays.
(b) Digital (radial) grooves.
(c) Handplate.

47 (a) The proximal limb passes behind the superior mesenteric artery.
(b) Its connection to the intestine shrinks and breaks.
(c) Proximal portion of the yolk stalk.

48 (a) Septum primum.
(b) Foramen ovale.
(c) Septum secundum.

49 (a) The anterior two-thirds of the tongue (oral part) forms from a median tongue bud (tuberculum impar) which fuses in the floor of the pharynx with two distal tongue buds (lateral lingual swellings) from each first branchial arch. The posterior third (pharyngeal part) forms from the fusion of the ventromedial ends of the second arches to form the copula. Caudal to the copula, a large hypobranchial eminence forms from third and fourth arch mesenchyme and overgrows the copula.
(b) Sulcus terminalis and foramen caecum.
(c) The two lateral swellings overgrow and bury the tuberculum impar and fuse to form the median sulcus.

50 (a) Exstrophy of the bladder.
(b) Defective closure of the anterior abdominal wall as a result of the lack of

muscle formation. The anterior wall of the bladder later ruptures as does the thin epidermis.
(c) No, 1 in 50,000 births, occurring more often in males.

51 (a) Di-, tri-, quadri, zygotic etc. or fraternal: this is the most frequent type of twins.
(b) Monozygotic or identical.
(c) Dizygotic twins normally have 2 amnions, 2 chorions, and 2 placentae. Monozygotic twins' membranes depend upon the stage at which twinning occurred, but they commonly have 2 amnions, 1 chorion, and 1 placenta.

52 (a) Pleuropericardial membrane.
(b) Thoracic wall.
(c) Common cardinal vein.

53 (a) The spinal cord grows more slowly than the dura mater and vertebral column.
(b) Yes, the spinal nerves run obliquely toward the intervertebral foramina to leave the vertebral column.
(c) Week 17-20 and continues until one year of age.

54 (a) Cranial portion.
(b) The mesonephric ducts, but their mesodermal epithelium is replaced by urogenital sinus endodermal epithelium.
(c) Median umbilical ligament extending from the apex of the bladder to the umbilicus.

55 (a) Elastic resistance of the lungs, surface tension of fluid in the lungs.
(b) Those nearest to the bronchi. Those at the periphery expand by three-four days of postnatal development.
(c) Hyaline membrane disease is a common cause of death in infants born before Week 32.
(d) Thyroxine.

56 (a) The first (mandibular) and second (hyoid) arches around the first branchial groove.
(b) Tragus.

57 (a) Secondary canalisation.
(b) Neural crest.
(c) There are no fixed curves.

58 (a) Cell axons growing out from the ventral horn of grey matter to each somite.
(b) Segmental aggregations of neural crest cells.

59 (a) From somatic mesenchyme *in situ*.
(b) Splanchnic mesenchyme.
(c) Dorsal and ventral primary rami.

60 (a) The sublingual develops from endoderm; the submandibular is also derived from endoderm; the parotid is derived from ectoderm lining the mouth.

(b) Outgrowths form branches and acini; these develop lumina and form the main ducts. Mesenchyme forms the remaining gland, except epithelium lining the ducts and acini.
(c) Parotid at Week 6-7; submandibular late in Week 6; sublingual at Week 8.

61 (a) Most of the epithelium and glands of the gut tube.
(b) Yolk sac remnant (stalk).
(c) It detaches from the gut and by Week 12 it is shrunken and hardened.

62 (a) Mesenchymal folds covered with ectoderm appear above and below the lens placode and grow toward each other.
(b) Tarsal plate and musculature.
(c) Tarsal glands and eyelashes.
(d) Week 9-10 and remain fused until Week 25-26.

63 (a) Cranial part of the gubernaculum.
(b) Paramesonephric duct.
(c) The ovary descends from the posterior abdominal wall to a point just inferior to the pelvic brim by Week 12.

64 (a) Parenchyma, acini and islets of Langerhans.
(b) Septa and connective tissue covering of the gland.
(c) Salivary glands.

65 (a) From the sheet of mesenchyme present between the two merged medial nasal prominences.
(b) Ectodermal. Their nerve fibres grow through the cribriform plate of the ethmoid into the olfactory bulb.
(c) Primarily through the nose. It only breathes through the mouth when nasal obstruction produces stress.

66 (a) Relatively common; particularly ventricular septal defects.
(b) Membranous.
(c) Days 20-50.
(d) Pulmonary stenosis, ventricular septal defects, overriding aorta and hypertrophy of the right ventricle.

67 (a) Originally ventrally and then rotates antero-medially.
(b) From Week 8. It is excreted into the amniotic fluid.
(c) The pelvic kidney is supplied by the iliac arteries. As it migrates it is supplied successively by new aortic branches. At Week 9 it meets the most caudal suprarenal artery from the aorta and acquires it to form the renal artery.

68 (a) 6-8 years of age. Their suture is rarely present in the adult.
(b) They are absent at birth, invade the bones at about 2 years, and are visible on X-ray at 7 years.
(c) It is depressed in severe malnutrition and dehydration and raised in hydrocephalic infants.

69 (a) It arises as a solid endodermal outgrowth from the liver diverticulum. It enlarges, canalises and forms a sac.
(b) Cystic duct.
(c) Bile pigment is secreted in Week 13-16 and colours the contents of the colon (meconium) green.

70 (a) Somites which are of mesenchymal origin.
(b) Skeletal muscles, bones, cartilage and dermal layer of the skin.
(c) Branchial arch myoblasts (except tongue and eye muscles).

71 (a) Reduction occurs in Week 10.
(b) The proximal limb forms the duodeno-jejunal loop and upper and middle parts of the ileum. The distal limb forms the caudal end of ileum, caecum, appendix and ascending and transverse colon.
(c) All except that of the midgut duodenum and ascending colon. The duodenum (except the first 25mm in the adult) and ascending colon are retroperitoneal.

72 (a) Dorsal aspect at the distal end of a digit.
(b) Eponychium.
(c) Cuticle.

73 (a) Efferent ductules of the testis and paradidymis.
(b) Epoophoron and paroophoron.
(c) It is suppressed and forms the longitudinal duct of the epoophoron, appendix vesiculosa and duct of Gartner.

74 (a) The basal part; areas painted dark blue, i.e. occipital bone, body and lesser wings of sphenoid, body of the ethmoid, otic and nasal capsules, petrous and mastoid parts of the temporals.
(b) Neurocranium is the part which protects the brain and viscerocranium forms the jaws.
(c) Maxillary process of branchial arch I: maxilla, zygomatic, and squamous parts of temporal, vomer, and palatine. Mandible forms around Meckel's cartilage (first arch ventral end).

75 (a) Stratum germinativum.
(b) Lanugo and vellus.
(c) Buds from a hair follicle root sheath.
(d) The glans penis and labia minora where they bud from the epidermis.

76 (a) Blood islands appear in the yolk sac and body stalk. The central cells form nucleated primitive red blood cells. The peripheral cells form the vascular endothelium of the blood vessels. Later several islands join to form a vessel.
(b) Aorta.
(c) Third aortic arches.

77 (a) Gonadotrophins, Week 9; growth and lactogenic hormones, Week 10; adrenocorticotrophic hormone, Week 8-9; thyrotrophic hormone, Week 19.
(b) Acidophilic granules appear at the beginning of the third month while basophilic granules appear from 3.5-4 months.

78 (a) Deciduous. The permanent teeth buds form lingual to the deciduous teeth during Week 10.
(b) Bud, cap and bell stages.
(c) Cap stage. Germ mesenchyme deep to the enamel organ condenses to form a dental papilla which will give rise to dentine and pulp.

79 (a) Myotomes originally on the dorsal (back) surface migrate in two sheets around the body to the ventral surface and then fuse in the midline.
(b) Linea alba.

80 (a) Cytotrophoblastic extensions from the villi penetrate the syncytiotrophoblast layer and join to form the cytotrophoblastic shell which anchors the two.
(b) As the chorion laeve forms, the villi in the basalis region increase to form the chorion frondosum. The basalis region forms the maternal component (decidua plate). As villi invade this area they divide the tissue into 10-38 cotyledons. Each contains two or more main stem villi,
(c) It extends into the uterine cavity until it contacts the decidua parietalis and fuses with it. By Week 22 it degenerates due to a reduced blood supply.

81 (a) Commonly bilobar but occasionally trilobar or alobulated.
(b) Two bilateral masses from the third ventral pharyngeal pouches which fuse in the midline.
(c) Endoderm gives rise to Hassall's corpuscles and cytoreticulum. Mesenchyme gives rise to the incomplete septa.
(d) After Week 10 they 'seed' the spleen, lymph nodes and Peyer's patches.

82 (a) Neuroblasts in the alar plate (laminae) migrate into the tectum.
(b) Basal plate.

83 Epithelium (endoderm) of the urogenital sinus (except the glandular part).
(b) An ectodermal glandular plate ingrows at the tip of the penis and then splits to form a urethral groove.
(c) Entirely from endoderm of the vesical part of the urogenital sinus.

84 (a) Endodermal derivative of the first pouch.
(b) The tubotympanic recess gives rise to the tympanic cavity. The proximal end forms the auditory (or pharyngotympanic or Eustachian) tube; the distal one expands to contribute to the tympanic cavity.
(c) Malleus, incus, stapes, their ligaments and the chorda tympani nerve.

85 (a) Scrotum.
(b) The part of the prostatic urethra that lies proximal to the opening of the prostatic utricle.
(c) It forms, but regresses.

86 (a) Initially the amnion's cellular walls. Later the majority from maternal blood, and in late pregnancy 500ml of fetal urine is excreted daily into the cavity.
(b) Physically cushions the embryo, maintains a constant temperature, prevents the amnion from adhering to the developing embryo and permits symmetrical growth and free movements of the fetus.
(c) Desquamated fetal epithelial cells, lanugo, vernix caseosa, proteins, fats, hormones, carbohydrates, enzymes, pigments, fetal urine and 98-99 per cent water.

87 (a) From lymph nodules in the root of the tongue.
(b) From the second pharyngeal dorsal pouch. They atrophy between 5 years and puberty.
(c) From lymph nodules around the auditory (Eustachian) tubes.
(d) The adenoids arise from lymph nodules in the nasopharynx. They are maximum in size at 6 years and involute by puberty.

88 (a) Week 7.
(b) No, true peristalsis occurs after birth.
(c) Week 18.

89 (a) Most of these malformations are uncommon.
(b) Treacher Collins syndrome and Pierre Robin syndrome.
(c) Failure of cranial neural crest cells to migrate in sufficient numbers into the first branchial arch.

90 (a) Canalicular, 16-25 weeks.
(b) End of the first trimester.
(c) Aspiration of amniotic fluid into the lungs.

91 (a) End of the eighth week.
(b) Right horn of the sinus venosus and the part intervening between the right and left horns.
(c) Left horn of the sinus venosus.

92 (a) Breech.
(b) Ultrasound.
(c) Radiation affects the developing fetus, e.g. leukaemia later in life.

93 (a) Week 3.5-8.
(b) Oligohydramnios.
(c) Thalidomide.

94 (a) Poorly developed taenia coli, the external surface is smooth, because it lacks haustra (sacculations) and appendices epiploicae.
(b) Ingested amniotic fluid, sloughed skin epithelial cells, oral cavity cells, upper respiratory and intestinal tract cells, lanugo hairs, vernix caseosa, secretions of the liver, pancreas and gastrointestinal glands, urea, steroids, biliverdin, mucoproteins and mucopolysaccharides.
(c) Meconium.
(d) Fetal distress, except late in labour in a breech presentation.

95 (a) Cavitation.
(b) Subendocardial tissue around the atrioventricular canals. These then become hollowed out on the caudal surfaces.
(c) Interatrial septum.
(d) Right, because it works harder in the fetus than the left. One month after birth the reverse is true.

96 (a) Cells from the rhombic lip and dorsal part of the alar lamina of the metencephalon.
(b) Pons and medulla.

97 (a) The liver is an endodermal derivative of the ventral foregut. The hepatic diverticulum penetrates the splanchnic mesenchyme of the septum transversum and divides into right and left branches. Columns of endodermal cells (liver cords) grow out into the mesenchyme.
(b) Connective, haemopoietic tissue, Kupffer cells and fibrous capsule of the liver.
(c) As the yolk sac regresses the liver enlarges and haemopoiesis occurs at Week 6, peaks between Week12-24 and ceases at birth.

98 (a) Suprarenal gland.
(b) Sympathetic ganglia.
(c) By the end of the third year.

99, 100 (a) Bladder.
(b) Male.
(c) Oligohydramnios.

101 (a) Approximately 1 in 5,000, a common congenital anomaly.
(b) Perineum or urinary tract.
(c) Lumbosacral spinal abnormalities, associated neural defects, urological abnormalities and oligohydramnios.

102 (a) Blastocyst (Stage 5).
(b) Ectoderm and endoderm.
(c) Ectoderm.

103 (a) Pseudoglandular.
(b) Visceral pleura form from splanchnic mesenchyme and parietal pleura form from somatic mesenchyme.
(c) Splanchnic mesenchyme.

104 (a) Rete testis, seminiferous tubules, tubuli recti.
(b) Mesorchium.
(c) Supporting Sertoli cells derived from germinal epithelium and spermatogonia derived from primordial germ cells.

105 (a) The apical ectodermal ridge influences growth of the limb.
(b) Skin, fingernails, hair and sweat and sebaceous glands.
(c) Muscles, bones, joints, blood vessels and lymphatics.

106 (a) Suprarenal (adrenal) gland.
(b) Medulla derives from neuroectoderm and cortex from mesenchyme.
(c) The medullary mass which forms the chromaffin cells is gradually surrounded by the fetal cortex. An additional layer of mesenchyme invests the suprarenal to form the adult cortex.
(d) Female pseudohermaphroditism.

107 (a) Cytotrophoblast and syncytiotrophoblast.
(b) They form when the primary villi acquire connective tissue cores.
(c) They form when capillaries develop in the secondary villi.

108 (a) Foramen caecum.
(b) During migration it is drawn out from its origin on the tongue as the thyroglossal duct. Its distal end forms the pyramidal lobe. The remaining duct normally disappears in Week 5.
(c) Aberrant thyroid tissue, sinuses, cysts and fistulae.

109, 110, 111 (a) Congenital dislocation of the hip.
(b) 1 in 1,500 births.
(c) Abnormal development of the acetabulum and generalised laxity of the ligaments of the hip joint.

112 (a) Week 11.
(b) Fourth ventral pouch.

(c) The parafollicular or 'C' cells which secrete and store calcitonin to regulate calcium levels in the body fluids.

113 (a) Mesenchyme surrounding the optic cup.
(b) Mesenchyme immediately anterior to the developing lens gives rise to the membrane. The centre of the membrane degenerates to form the pupil.
(c) Buds from the surface ectoderm branch, canalise and form ducts and alveoli.
(d) An ectodermal cord along a line where the lateral nasal prominence and maxillary prominence meet. The cord later canalises.

114 (a) Cryptorchidism.
(b) Week 28-32. Over 97 per cent of male neonates have both testes in the scrotum.
(c) The gubernaculum on the lower side of the testis lies obliquely in the abdominal wall and attaches to the scrotal swelling. As the fetal body elongates, the gubernaculum grows at a slower rate so that the testis descends into the scrotum.
(d) A sac of parietal peritoneum carrying with it layers of the abdominal wall, which form the coverings of the spermatic cord and testis.

115 (a) Scrotal raphe which is continuous with the perineal raphe.
(b) Mesenchyme in the phallus.
(c) An epithelial fold ingrows from the tip. After birth, this fold splits to form the inner epithelial lining of the prepuce and the outer epithelial lining of the glans.

116 (a) Week 7.
(b) Primitive mediastinum (mesenchyme ventral to the oesophagus).
(c) Right. The membrane may be larger on this side due to the larger right common cardinal vein.

117, 118 (a) Dilated.
(b) Yes.
(c) 1000ml. The fetus swallows and absorbs about 400ml per day.
(d) Every three hours.

119 (a) Central tendon.
(b) Pleuroperitoneal membranes, dorsal mesentery of oesophagus and body wall.
(c) Cervical 3, 4 and 5.

120 (a) Right atrium; the sinus venosus then takes over.
(b) Dilation of the ventricles as the heart grows in length.

121 (a) The ascending aorta is derived from the left half of the truncus arteriosus. The aortic arch is derived from the aortic sac and the left fourth branchial arch artery. The descending aorta is derived from the left dorsal aorta.
(b) Proximal parts of the right and left sixth arch arteries.
(c) Distal part of the left sixth arch artery.

122 (a) Thymus gland.
(b) Right sinoatrial valves.
(c) Sinoatrial valves fuse in their upper part.

123 (a) An area of skin supplied mainly by a single spinal nerve and its dorsal root ganglion.
(b) No sensory overlap.
(c) The palmar dermatome is mainly C7; C6 and C8.
(d) The plantar dermatome is mainly L5; L4 and S1.

124 (a) Week 25 and 26.
(b) Lateral and central sulci.
(c) Brain surface area is increased relative to volume.

125 (a) Two mandibular prominences, two maxillary prominences and the frontonasal prominence.
(b) Stomodeum.
(c) Rudimentary upper and lower jaws, unerupted teeth and small maxillary and nasal sinuses.

126 (a) Distal part of the transverse colon, descending and sigmoid colon, rectum, upper part of the anal canal and part of the urogenital system.
(b) Inferior mesenteric artery.
(c) The mesentery of the descending colon fuses with peritoneum on the left dorsal abdominal wall and disappears. The mesentery of the sigmoid colon is retained; the mesentery of the rectum disappears.

127 (a) Occipital myotomes.
(b) Week 7-9.
(c) Sensory: lingual branch of the mandibular division of V. Taste buds (except vallate): chorda tympani of VII. Vallate papillae: IX. Muscles: hypoglossal.

128 (a) Amnion, chorion, allantois and yolk sac.
(b) Fetal chorion and maternal endometrium.
(c) Protects the embryo or fetus and provides for nutrition, respiration and excretion and produces hormones.

129 (a) Myotome region of somites.
(b) Iris muscles form from neuroectoderm.

130 (a) Greater omentum.
(b) Epiploic foramen.

131 (a) Hepatic sinusoids form as liver cords anastomose enclosing endothelium-lined spaces present in the septum transversum.
(b) Week 12.
(c) Week 10-12.

132 (a) The midbrain flexure during Week 4, the pontine flexure in Week 5 and the cervical flexure during Week 5-7.
(b) The cervical flexure between the hindbrain and spinal cord.
(c) Pontine.

133 (a) Spina bifida.
(b) Failure of fusion of the vertebral arches and malformation involving the caudal end of the neural tube.

134 (a) Week 24.
(b) Type II.
(c) To reduce the surface tension of the fluids lining the airways.
(d) The majority form after birth until 8 years of age.

135 (a) Normal stump regression may be used to determine infant survival time.
(b) When blood flow is interrupted they gradually become fibrous cords. The arteries remain patent at the proximal end where a superior vesical branch supplies the bladder.
(c) Left umbilical vein.

136 (a) Urethral groove.
(b) Urethral groove and urogenital orifice.
(c) A median penile raphe over the tubular urethra.

137 (a) Malleus and incus.
(b) Stapes.
(c) Tensor tympani from first arch and stapedius from second arch.
(d) Mastoid or tympanic antrum.

138 (a) A dilatation of the foregut.
(b) Ventral surface.
(c) Inferior end of oesophagus, stomach, superior end of duodenum.

139 (a) Bulbar ridges of the bulbus cordis.
(b) Aorticopulmonary septum.

140 (a) Ectodermal neural plate.
(b) Week 2-3.
(c) Neuroectoderm forms neural folds which rise and fuse in the midbrain region to form a tube.

141 (a) The lower part of the otocyst (cochlear) division.
(b) Utriculosaccular duct.
(c) Three circular pouches arise from the utricular part; their central walls fuse and are resorbed. The peripheral walls remain as the semicircular ducts.

142 (a) Week 9.
(b) Glans penis or clitoris.
(c) The two urogenital folds enclose the membrane and the two labioscrotal swellings enclose the urogenital folds.

143 (a) Stage 5, Days 7-10 after fertilisation.
(b) Reactions of the uterine endothelium to implantation.
(c) Decidua capsularis is the area overlying the conceptus; decidua basalis is the area underlying the conceptus; decidua parietalis is the remainder of the maternal mucosa.

144 (a) The premaxillary part of the maxilla forms in membrane. The maxillae and palatine bones extend into the lateral palatine processes to form the hard palate.
(b) Soft palate.

145 (a) Tympanic membrane.
(b) Year 2.
(c) The facial nerve emerging from the stylomastoid foramen is exposed and relatively unprotected.

146 (a) Neural crest cells migrate to form the spinal ganglion. Processes from these neurons grow in a single bundle into the spinal cord opposite the apex of the dorsal horn of grey matter.
(b) Motor apparatus (ventral or anterior horn).
(c) Lower level of L2—upper level of L3.

147 (a) Cranial is preaxial; caudal is postaxial.
(b) The upper limb has a better blood supply than the lower limb.
(c) It moves from 90^0 when the arm, forearm, and hand form, so that the palm faces the midline. At Week 7-9 the elbow is moved dorsally.

148 (a) Aorta and pulmonary trunk.
(b) Right: conus arteriosus; left: aortic vestibule.

149 (a) They are present but are concealed by a thick fat pad.
(b) Week 17.

150 (a) Glandular epithelium forms from numerous outgrowths of the prostatic urethral endoderm and the stroma and smooth muscle form from surrounding mesenchyme.
(b) It forms as an endodermal outgrowth of the membranous urethra and the stroma and smooth muscle form from mesenchyme.

151 (a) Endoderm.
(b) Mesoderm which forms vascular blood, and primordial germ cells.
(c) Provides nutrients while the chorioallantoic placenta develops.

152 (a) It occupies most of the abdominal cavity.
(b) Haemopoiesis.

153 (a) Two endodermal buds from the foregut.
(b) Week 20.
(c) Duodenum.

154 (a) Quickening.
(b) Approximately 9-10 weeks.
(c) At their angles.

155 (a) Flat bones, i.e. frontals, parietals, squamous parts of the temporals, nasal and lacrimal bones.
(b) New bone is deposited on the external surface and there is destruction of old bone on the internal surface. Peripheral spread is by ossification occurring in the intervening mesenchymal tissue.

156 (a) Yes. The oesophagus has a ciliated lining at Week 11-28.
(b) Before Week 8 when it recanalises.
(c) Striated muscle originates from the caudal branchial arches mesenchyme. Non-striated muscle develops from splanchnic mesenchyme. Both types are supplied by branches of the vagus nerve.

157 (a) Week 6-10.
(b) An ectodermal thickening, the dental lamina, grows into the boundary process. A second lamina grows in between the dental lamina and the future lip. As the teeth develop, the labiogingival lamina degenerates leaving a deepening groove separating the lips and cheeks from the teeth and gums.

158 (a) Left umbilical vein.
(b) Ligamentum teres.
(c) It undergoes atrophy and disappears at 6-7 mm CR (Stage 13). As a result all blood from the placenta enters through the left umbilical vein.

159 (a) An indentation on the inferior aspect of the optic stalk.
(b) Vascular mesenchyme enters the choroidal fissures, then blood vessels form and grow from the proximal to the distal optic cup. In Week 7 the optic fissure fuses and the blood vessels are incorporated into the optic stalk.
(c) The hyaloid artery and vein which pass from fissure to lens.

160, 161 (a) The absence of forebrain neural tissue. This is usually confirmed by the lack of the optic nerves.
(b) Yes, as early as Week 14.
(c) Acrania: absence of the cranial vault.

162 (a) An hernia is covered by subcutaneous tissue and skin, whereas an omphalocele is covered with peritoneum and amnion.
(b) Omphalocele.

163 (a) The caudal end of the cerebral hemisphere turns ventrally and then rostrally to bury the insula and form the temporal lobe.
(b) The two cerebral hemispheres grow, meet in the midline and, as they flatten medially, mesoderm is trapped between them which forms the falx cerebri.

164 (a) From the left segments (Week 6).
(b) Variations are common but anomalies are rare.
(c) Lesser omentum (hepatogastric ligament and hepatoduodenal ligament), falciform ligament, and visceral peritoneum of the liver.

165 (a) Forehead, bridge and apex of the nose.
(b) Nasal sac.
(c) The caudal part of the nasal fin (oronasal membrane) breaks down to connect the nasal and oral cavities.

166 (a) Ventricular (ependymal); intermediate (mantle); marginal.
(b) Cortex. Cells from the mantle layer migrate into the marginal layer to form the cortex.

167 (a) Ductus arteriosus.
(b) It constricts and is functionally closed by 10-15 hours.
(c) Bradykinin.

168 (a) Dorsal mesocardium.
(b) Transverse pericardial sinus.
(c) The central part of the dorsal mesocardium degenerates to form an opening.

169 (a) Prosencephalon (telencephalon, diencephalon); mesencephalon; rhombencephalon (metencephalon, myelencephalon).
(b) Cerebral hemispheres.
(c) Lamina terminalis.

170 (a) The anal canal epithelium becomes keratinised.
(b) Anal epithelium changes from columnar to stratified squamous epithelium.
(c) Anal membrane.

171 (a) This is the fetal side which is smooth with a central umbilical cord. The maternal side is raised into 10-38 cotyledons.
(b) Metabolism and transfer of nutrients, wastes, antibodies, hormones and electrolytes. It also supplies hormones.

172 (a) Enamel is first formed in the cusp and then progresses toward the root of the tooth.
(b) Pre-dentine forming odontoblasts.
(c) Odontoblast processes remaining in the dentine.

173 (a) Median palatine process (primary palate); two lateral palatine processes (secondary palate).
(b) Week 7-12.
(c) The two lateral processes which originally project vertically downwards on either side of the tongue meet and fuse in the midline. They also fuse with the primary palate and nasal septum.

174 (a) The superior glands develop from endoderm of the fourth pharyngeal dorsal pouch. The inferior glands develop from third pharyngeal dorsal pouch endoderm.
(b) The third pharyngeal pouch gives rise to the thymus gland and inferior parathyroid glands. The fourth pharyngeal pouch gives rise to the superior parathyroids. All parathyroids migrate caudally to their final positions on the posterior thyroid surface.
(c) Week 12.

175 (a) Week 16-20.
(b) It is used to establish an approximate age for medico-legal purposes.
(c) Yes. At birth all bones in the skeleton have centres of haemopoiesis (red marrow).

176 (a) The head rotates to natural position relative to the shoulders.
(b) Three stages. First, onset of labour until complete dilation of the cervix; second, complete dilation of the cervix to delivery of the fetus; third, delivery of the fetus to delivery of the placenta and membranes.

177 (a) About Week 27.
(b) White (yellow) and brown.
(c) Root of the neck, behind the sternum, perirenal area, around organs in the thorax and posterior abdominal wall. It is important in metabolism and heat production.

178 (a) Endoderm.
(b) Branchial membrane.
(c) It is rudimentary or absent.

179 (a) Fourth-fifth month.
(b) Toward the end of the fourth month.
(c) They are poorly developed and the intestine has a very thin wall.

180 (a) The anterior wall forms cuboidal anterior lens epithelium. The posterior wall hypertrophies and forms lens fibres.
(b) They degenerate distally and disappear after Week 31. Their proximal parts form the central artery and vein of the retina.

181 (a) 10 to 12 per 10,000 individuals aged between birth and 5 years. In associated congenital heart anomalies the rate increases to 20 per cent.
(b) Trisomy 21 (Down's) Syndrome.
(c) 20 per cent in Trisomy 21, otherwise it is a relatively rare condition.

182 (a) Mesenchyme proliferates around the anal membrane causing the pit to form. The membrane ruptures in Week 8 and the anal canal communicates with the amniotic cavity.
(b) Perineal body.
(c) Meconium.

183 (a) Second dorsal pouch.
(b) Third ventral pouch.
(c) Obliterated by tongue formation.

184, 185 (a) Dwarfism.
(b) That of the largest fetus.
(c) Head and abdominal circumference.
(d) It is an established cause of intrauterine growth retardation.

186 (a) A collection of white fat over the buccinator muscle.
(b) It prevents the cheeks being drawn inwards during sucking.
(c) There is a lack of subcutaneous fat and the skin grows faster than the underlying connective tissue.

187 (a) Roof of the metencephalon.
(b) Floor of the metencephalon.
(c) It lies caudal to the acousticofacial mass and caudal to the widest part of the diamond-shaped hindbrain.

188 (a) Neural crest.
(b) Dorsal roots of spinal nerves.

189 (a) On their medial aspect.
(b) They atrophy until only those at L1-L3 remain.
(c) It forms the appendix of the epididymis, duct of the epididymis, vas (ductus) deferens and ejaculatory duct.

190 (a) Intermaxillary segment.
(b) Philtrum of the upper lip, premaxillary part of the maxilla and its gingiva (gum) and the primary palate.
(c) Alae.

191 (a) Fetal cortex.
(b) Neural crest.
(c) The cortex produces corticosteroids, androgens, and oestriol precursors. The medulla produces insignificant amounts of epinephrine (adrenalin).

192 (a) Endoderm.
(b) Oropharyngeal membrane (buccopharyngeal membrane).
(c) Primitive gut and amniotic cavity.

193 Upper two-thirds: inferior mesenteric artery; lower one-third: internal pudendal artery.

194 (a) Superior mesenteric artery.
(b) The yolk stalk divides the loop into cranial and caudal limbs. The caecal diverticulum separates small and large intestine.
(c) The abdomen is mainly occupied by the liver and kidneys and the rapidly growing gut becomes extraembryonic.

195 (a) Two pairs develop: the jugular and the iliac. Two single sacs develop: the retroperitoneal sac and the cisterna chyli.
(b) Vessels grow out from the sacs throughout the body.
(c) Connective tissue framework and capsule of lymph nodes.

196 (a) First arch is nerve V or trigeminal; second arch is nerve VII or facial; third arch is nerve IX or glossopharyngeal.
(b) Anterior ligament of malleus, sphenomandibular ligament.
(c) Stapedius, stylohyoid, posterior belly of digastric, buccinator, auricularis, frontalis, platysma, orbicularis oris and oculi etc.

197 (a) Crista dividens and valve of the inferior vena cava.
(b) Right anterior cardinal and right common cardinal veins.
(c) Septum primum.

198 Fistula—an abnormal canal, drawing 2; atresia—absence of a lumen, drawing 1; stenosis—narrowing of a lumen, drawing 3.

199 (a) The dorsal (alar) and ventral (basal) laminae.
(b) Four colliculi, red nucleus, nuclei of cranial nerves III and IV and reticular nuclei.

200 (a) Six pairs of arches, each separated from the next by a branchial groove or cleft. The last two arches are vestigial and the fifth arch may be entirely absent.
(b) Cranial part of the foregut.
(c) Cervical sinus.

201 (a) First branchial groove.
(b) First branchial membrane.
(c) Meatal plug ectoderm, first and second branchial arch mesenchyme and tubotympanic recess endoderm.

202 (a) The lungs have sunk. This is because the infant was stillborn and the lungs contain fluid and not air.
(b) Amniotic fluid and liquid from the tracheal glands and lungs.
(c) Pressure on the thorax during delivery forces fluid through the nose and mouth; by pulmonary capillaries; by lymphatics and pulmonary vessels.
(d) Pressure in the left atrium markedly increases over that in the right atrium; the foramen ovale is closed as the septum primum presses against the septum secundum.

203, 204 (a) Congenital hydrocephalus.
(b) Ventricular hemisphere ratios, not the presence or absence of a large head.

205—208 (a) B, D, C, A.
(b) Between Week 7-9.
(c) The fetal position: the limb is flexed, abducted at the hip joint, the knee flexed and the foot inverted in the talipes varus position.

Acknowledgements

In addition to materials I have prepared for this book, I am very grateful to many individuals and Institutions for allowing me to photograph specimens.

Professor T.W.A. Glenister, C.B.E., and the Anatomy Department, Charing Cross and Westminster Hospital Medical School kindly allowed me to study their collection. Miss Marion Hudson, Medical Illustration, generously advised me and photographed the embryos.

I am also deeply indebted to Professor K.E. Carr and the Department of Anatomy for the privilege of allowing me to present some of the materials in The Belfast Collection, The Queen's University, Belfast. Mr G. Bryan, Chief Technician, gave of his time and expertise.

Dr E.C. Blenkinsopp, Consultant Pathologist, and Dr T. El-Sayed, Consultant Radiologist, Watford General Hospital, very kindly helped me and without their assistance much of my work would be impossible.

I should especially like to thank Mr K. Garfield, Chief Technician, Leicester University Central Photographic Unit for his cheerfulness, understanding and creativity during the preparation of this book. The remaining photography is my own and should not in any way reflect upon his expertise.

I am also very grateful to the following people for their generosity: Professor H. Nishimura, Professor Emeritus of Anatomy, Kyoto University; Dr P. McKeever, Senior Lecturer in Paediatric Pathology and Dr R. Walker, Senior Lecturer, Department of Pathology, Leicester Medical School; Group Captain C.E. Ockleford Q.H.D.S., Oral Surgeon, R.A.F. Hospital, Akrotiri, Cyprus; Mr John Grenfell and the Medical Illustration Department, Leicester Royal Infirmary; Mr P.A. Runicles, Chief Technician, Department of Anatomy, Middlesex Hospital Medical School; Mr G.L.C. McTurk, Leicester University Scanning Electron Microscope Unit; Mr C.G. Brooks, Mr H.J. Kowalski, Mr I. Patterson, Leicester University Central Photographic Unit; Mr D. Adams, Department of Anatomy, Leicester Medical School; The Royal College of Obstetricians and Gynaecologists; and the Anatomy Department, The London Hospital Medical College.

I am deeply indebted to Professor Angus d'A. Bellairs, St Mary's Hospital Medical School, London; Mr G.C. Tressider, Surgeon and Anatomist and Mr B. Benninger, Leicester University Medical School who generously gave their time and read the manuscript. Dr J.M. England, Consultant Haematologist, Watford General Hospital also read the manuscript and compiled the index.

Index

Figures in **bold** type refer to illustrations; those in light type refer to pages.

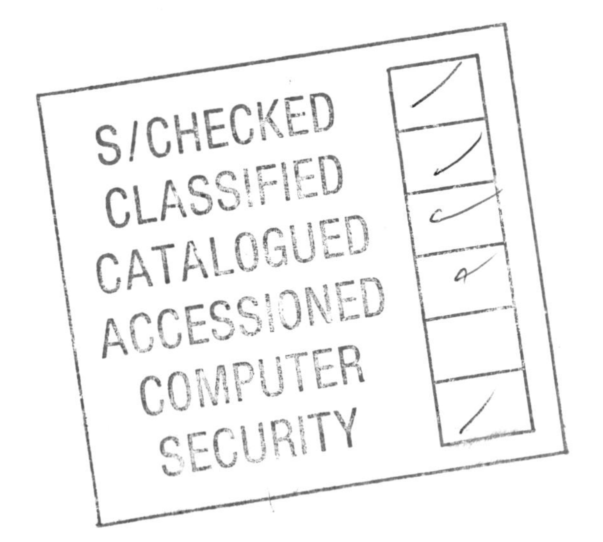